CW01072729

DOCKS AND KN
OF
BRITAIN AND IRELAND

A second edition of
DOCKS AND KNOTWEEDS
OF THE BRITISH ISLES
by J.E. LOUSLEY and D.H. KENT

B.S.B.I. HANDBOOK No. 3

J.R. AKEROYD

ILLUSTRATED BY ANN FARRER

EDITED BY P.H. OSWALD & J.R. EDMONDSON

BOTANICAL SOCIETY OF BRITAIN AND IRELAND
London
2014

First published (as *Docks and Knotweeds of the British Isles*) 1981

Second edition (*Docks and Knotweeds of Britain and Ireland*) 2014

ISBN 978-0-90-115847-5

Published by the Botanical Society of Britain and Ireland
c/o The Natural History Museum
Cromwell Road, London SW7 5BD, UK

Cover design by Annunziata Connolly
Printed by Henry Ling Limited, Dorchester

CONTENTS

PREFACE TO THE SECOND EDITION

Several important developments in the taxonomy of the Polygonaceae have taken place since the publication of *Docks and Knotweeds of the British Isles* in 1981. A new account of the family appeared in a revised first volume of *Flora Europaea* (Tutin *et al.* 1993), a new British Flora has been published (latest edition, Stace 2010) containing a significantly altered account of the family, and there is now widespread recognition of the necessity to divide *Polygonum* sensu lato into segregate genera. The taxonomy of eastern Asian Polygonaceae has been substantially reviewed in the English edition of *Flora of China* (Li *et al.* 2003), and a full account of the species in the USA and Canada is available in *Flora of North America* (Freeman & Reveal 2005). At the same time, continued study of the native and alien floras of Great Britain and Ireland has added to our knowledge of the distribution of rare and critical taxa in the family. This work and the stimulus of the original edition of the Docks and Knotweeds handbook have led to the discovery or rediscovery of rarer species, intraspecific taxa and hybrids (notably in *Rumex*) in British and Irish Polygonaceae. On the debit side, the phasing out of 'shoddy' from British agriculture and the decline in the indigenous wool industry have meant that fewer casual alien Polygonaceae now reach these islands.

I have been interested in Polygonaceae since 1975, when I began research towards my PhD on the genecology of *Rumex crispus*, Curled Dock. A workshop on the identification of Polygonaceae held at the University of Reading in August 1985, attended by more than 40 BSBI members, made me aware of the level of interest in the family amongst British botanists and gave me the impetus to revise the handbook in collaboration with the late Douglas Kent. My involvement with the revision of the first volume of *Flora Europaea* and the supervision of a MSc thesis by Louis Ronse Decraene on generic limits in *Polygonum* sensu lato, as well as contributing accounts of the family for *Mountain Flora of Greece*, *The European Garden Flora* and the BSBI's *Plant Crib 1998*, helped me synthesise the often scattered data on British and Irish Polygonaceae for this new edition. The subsequent delay in publication is entirely due to circumstances in my life, and I am grateful for the patience and support of friends and colleagues in the BSBI.

The descriptions from the first edition of this handbook remain substantially unchanged, although thoroughly checked and revised where appropriate. I have updated the taxonomy and nomenclature, following advice from the late Douglas Kent, and expanded the ecological and geographical information under each species. I have made some considerable changes to the introductory material, many of them necessitated by profound revisions of taxonomy and nomenclature within the family in recent years. In this new edition, *Persicaria* is recognised as a distinct genus embracing most of the species included within *Polygonum* in the first edition. *Fallopia* has been substantially revised to include *Reynoutria*, especially in view of the extensive studies of John Bailey, Ann Conolly and Michelle Hollingsworth. The accounts of *Rumex acetosella* (formerly referred to as '*Rumex acetosella* group'), *R. acetosa* and *R. crispus* have been rewritten to accommodate a substantial body of more recent bio-systematic and taxonomic research. The bibliography has been expanded and updated.

Sadly, I never knew Ted Lousley, although I possess one of his larger vascula (perhaps for gathering *Rumex*). I saw him in November 1975 at the very last BSBI Annual Exhibition Meeting that he attended, but, then a callow post-graduate student, I was too diffident to talk to him! It was, however, a great pleasure to work with his old friend and colleague Duggie Kent during the preparation of this new handbook, and I hope that they would have approved of the changes that I have made. Certainly, much of the research that followed, notably within the genus *Rumex*, owed a debt to their considerable efforts over many years.

June 2014
J.R.A.

ACKNOWLEDGMENTS

The author and editors are greatly indebted to the many specialists who have supplied information and advice on native and introduced species and hybrids of Polygonaceae. Foremost among these, as stated above, was the late Douglas Kent, co-author of the first edition of this handbook. The early stages of its revision were overseen by Philip Oswald, Handbooks Editor for the Botanical Society of the British Isles, to whom we are grateful for his meticulous and thorough work on points of format, terminology and taxonomy. We should also like to thank Ann Farrer, the artist of the plates published in the first edition (as Ann Davies), for not only consenting to our re-using them but also in supplying additional illustrations for this edition.

The author wishes to record his thanks to the numerous BSBI members, who over three decades have passed on so many interesting specimens of docks and knotweeds for information, checking or determination, and to Dr Stephen Jury and Ronald Rutherford for their help during his time working in the University of Reading herbarium (**RNG**) and on subsequent visits.

The distribution maps, prepared by Colin Handover and Steph Rorke, were produced by the Biological Records Centre, to whom we are most grateful.

Among the specialists who have read and commented on earlier drafts or supplied additional information, we wish to thank (in alphabetical order) John Bailey, David Beerling, the late Ann Conolly, John Parnell, Chris Preston, Louis Ronse Decraene, the late Maura Scannell, the late Peter Sell (who introduced the author to *Rumex*) and Clive Stace.

Finally, the long-suffering members of the Publications Committee of the Botanical Society of Britain and Ireland are to be thanked for their patience. We hope their wait will have been worthwhile.

PREFACE TO THE FIRST EDITION

Following the unqualified success of the first BSBI handbook, *British Sedges*, published in 1968, it was decided to sponsor further handbooks covering other families, genera or groups. One of those selected was the family Polygonaceae, to be written by J. Edward Lousley, who had for many years specialised in the study of the critical genus *Rumex*. A provisional timetable was drawn up in January 1975, and it was hoped that the plates and text would be completed by the end of that year.

When Ted Lousley suddenly died in January 1976, he had prepared draft accounts of most species of *Polygonum*, but sadly had completed less than half of the accounts of *Rumex*, including the many hybrids, all of which he knew so well. It was unfortunate that at the time there was no successor with a good knowledge of the genera involved who would have been able to write the outstanding accounts and finish the work. At a meeting of the BSBI Publications Committee early in 1976 the situation was reviewed, and I agreed, with the assistance of various other members of the committee, to attempt to complete the book. This I have now done, though regrettably my efforts have been restricted by very limited spare time and a long illness in the latter part of 1976.

I should like to express my thanks to the friends and colleagues who contributed notes and advice, or have read and commented upon the typescript. In particular I should like to thank Dr Norman Robson, who has seen the book through the press, and Arthur Chater; they have been towers of strength, and without their assistance the book might not have appeared. I am grateful also to Mrs Dorothy Lousley for making available the study books, manuscripts and specimens from her late husband's library and herbarium. Finally, I should like to thank Mrs Ann Davies for her co-operation and patience during the preparation of the illustrations.

November 1980
D.H.K.

INTRODUCTION

The primary object of this illustrated handbook is to assist in the identification of knotgrasses, knotweeds, persicarias, docks, sorrels and their relatives in Great Britain and Ireland. Although the Polygonaceae appear intractable to some botanists, many species are not critical and are easily identified. *Polygonum* and some species of *Persicaria* and *Rumex* can present difficulties to the non-specialist, largely owing to the occurrence of genetic and environmentally induced variation within certain taxa and, in *Rumex*, frequent hybridisation between species.

All native and introduced but established species, as well as most casual species (ephemeral introductions), are described and illustrated. Keys are provided to genera, subgenera, species and subspecies. Nomenclature (except for *Persicaria* and *Fallopia*) is based for the most part on that used in the second edition of Volume 1 of *Flora Europaea* (Tutin *et al.* 1993), and English names follow for the most part the third edition of *New Flora of the British Isles* (Stace 2010). Descriptions are followed by chromosome numbers, usually the earliest known count, with any British or Irish count denoted by an asterisk. After the chromosome numbers, data on flowering period are provided, though these are not given for casual adventive species. Then comes information on status, habitat, frequency, geographical distribution and taxonomic or ecotypic variation.

For this second edition of *Docks and Knotweeds*, hybrids are described under the account of one of the parent species, rather than segregated at the end of the book as in the first edition. The valid hybrid name is given first, followed by synonyms, where appropriate, and lastly the reference numbers and names of the parent species. Our knowledge of *Rumex* hybrids has expanded considerably over the last 30 years, with the discovery of several new to Great Britain and Ireland and some new to science, as well as numerous new distribution records.

In the case of many adventives, particularly those formerly introduced with wool-shoddy, it should be appreciated that in Great Britain or Ireland these can exhibit considerable variation in features such as plant size, growth and leaf shape relative to their appearance within their native range. A number have not been seen for several decades but they are included in case they turn up again with changing trends in trade and agriculture.

The handbook concludes with a glossary, references and indices to English and scientific names.

CLASSIFICATION

Polygonaceae is a family of some 50 genera and 1050 species, most of which occur in the temperate parts of the northern hemisphere. In Great Britain and Ireland, 51 species belonging to nine genera are native or well established and another 32 species, as well as two genera, have occurred only as casual adventives.

Species belonging to this family found in Great Britain and Ireland are easily recognised as such, since at each node they possess an ochrea (plural ochreae; sometimes spelled 'ocrea', 'ocreae') – a sheathing structure not unlike a miniature knee-bandage. These are formed by the fusion of two stipules, arising from the base of the petioles and embracing the stem. They are often very conspicuous and are an outstanding feature of the family, though the large North American genus *Eriogonum*, a few species of which are cultivated in rock-gardens, and some allied genera, do not have them. In Great Britain and Ireland most of the species in this family are herbaceous, the only species that are at all woody being the introduced *Fallopia baldschuanica* and *F. aubertii*, and *Muehlenbeckia complexa*. Carlquist (2003) presents a full review of the wood anatomy of Polygonaceae.

The flowers of Polygonaceae are generally hermaphrodite and small, and they are usually borne in large numbers in numerous or compound inflorescences. The perianth consists of two whorls of segments that are not differentiated into sepals and petals, i.e. they are 'tepals'. In *Rumex* and some other genera the basic number of tepals is three, and each flower has three outer and three inner ones. After fertilisation the inner tepals usually enlarge into structures termed valves and develop hooks, spines, wings or corky tubercles which play an important part in fruit-dispersal and provide valuable characters for classification and identification. The outer tepals show little change after fertilisation. The flowers of *Rumex* are generally green, the three stigmas are branched and feathery, and the anthers are borne on slender filaments easily shaken by the wind – characteristic adaptations to anemophily (wind pollination). The green developing inflorescence is also important in photosynthesis, with the perianth-segments in *R. crispus* recorded as constituting 69% of the plant's chlorophyll-containing tissues (Maun 1974).

In *Polygonum* and *Persicaria* the flowers are acyclic, with usually five tepals, which are often pink or reddish. This pattern probably arises from the fusion of one of the inner whorl of three tepals with one of the outer whorl, thus reducing the number from six to five. The stigmas are capitate, and nectar is secreted at the base of the stamens. These characters and the bright colour of the flowers in some species are associated with the attraction of insect visitors, in which the success of individual species varies considerably

The following general description applies to all the species of Polygonaceae found in Great Britain and Ireland:

Herbs, climbers or scrambling shrubs; stem usually leafy, often with swollen nodes; leaves mostly alternate, simple with characteristic sheathing stipules or ochreae that clasp the stem above the leaf-base. Inflorescence primarily racemose; the partial inflorescences usually cymose. Flowers usually hermaphrodite, regular, cyclic or acyclic. Perianth persistent, mostly forming a membranous wing round the fruit, biseriate or uniseriate with undifferentiated segments. Stamens (3–)6–9. Ovary superior, unilocular, with a single erect orthotropous ovule; styles 2–4. Fruit a trigonous, triquetrous or sometimes lenticular nut with a smooth or minutely punctulate surface. Seed containing an eccentric or straight embryo surrounded by mealy endosperm, sometimes ruminate.

TAXONOMIC LITERATURE

Most species of Polygonaceae in Great Britain and Ireland belong to a group of genera formerly included within *Polygonum* or to the larger genus *Rumex*. *Polygonum* sensu lato was revised in a monograph by Meisner (1826), when it included the genera *Fallopia* Adans. (as section *Tiniaria* Meisn.), *Persicaria* (L.) Mill. and *Reynoutria* Houtt. (as part of section *Aconogonon* Meisn.). The genus was later revised again by Meisner (1856). Much discussion of generic limits in *Polygonum* and related genera has taken place since that date. Most modern authors have recognised the need to separate *Persicaria* from *Polygonum* (*vide* Haraldson 1978, Ronse Decraene & Akeroyd 1988, Hong 1992, Ekman & Knutsson 1994), but further subdivision of *Persicaria* is more contentious, certainly among British botanists.

Several important treatments of *Polygonum* sensu lato have been published since that of Small (1895), who revised all the North American species, and Stewart (1930), who dealt with the species of eastern Asia, including a number that are grown in and naturalised from British and Irish gardens. Stearn (1969) and Akeroyd (1989) have provided keys to distinguish most of the species that are likely to be found in European gardens.

In particular, *Polygonum* sensu stricto (*Polygonum* section *Avicularia* Meisn.) has presented many difficulties to taxonomists. A detailed treatment was provided by Lindman (1912), although much of his work is no longer accepted. More recently the British species were monographed by Styles (1962), while Mertens & Raven (1965) and Wolf & McNeill (1986) provided useful accounts of North American species. Of several accounts of this section published in recent years, among the most useful are those of Raffaelli (1982), with descriptions and illustrations of the species present in Italy, Schmid (1983) for Bavaria, Meerts *et al*. (1983) for Belgium, and Karlsson (2000), who gives a detailed and incisive account of the taxa in Fennoscandia. Most of these also occur in Great Britain (though fewer in Ireland) as natives or aliens.

The taxonomy, cytology and genetics of *Fallopia*, now regarded by most authors as including *Reynoutria* (Ronse Decraene & Akeroyd 1988), has recently received detailed study (e.g. Bailey & Stace 1992, Beerling, Bailey & Conolly 1994, Hollingsworth & Bailey 2000).

The first detailed monograph of *Rumex* was that of Campderá (1819). Further accounts were produced by Murbeck (1899, 1913) and Danser (1916 *et seq*.). B.H. Danser studied *Rumex* and *Polygonum* in both the Netherlands and Dutch overseas territories (Danser 1927). Over seven decades K.H. Rechinger fil. (1906–1998) published a major series of monographs of *Rumex* worldwide, including regional accounts, which are of the utmost importance to students of the genus (Rechinger 1932 *et seq.*). It should be noted that Rechinger's studies on *Rumex*, with those of his father K. Rechinger (1867–1952), extended over more than a century! In Britain, Lousley (1939b, 1944 *et seq.*) made an extensive taxonomic investigation of the native and alien species, including studies of genetic variation and experiments on seed dispersal and germination. He was an efficient worker who wrote up all that he did, and most of his data on *Rumex* were incorporated into the first edition of this handbook.

Rumex is divided into four subgenera, recognised as distinct genera by some botanists, especially in eastern European countries. These subgenera are easily distinguished (see below), but convention, practicality and the wide usage of current names all weigh heavily against the use of separate generic names, which would only lead to confusion.

Other monographic treatments in Polygonaceae include studies of *Oxyria* (Emden 1929), *Koenigia* (Hedberg 1997) and *Fagopyrum* (Chen 1999).

ECOLOGY

Polygonaceae contains some of the commonest and most widespread of British and Irish weeds, both ruderals and weeds of cultivation, such as *Polygonum aviculare*, *Persicaria maculosa*, *P. lapathifolia*, *Fallopia convolvulus*, *Rumex crispus* and *R. obtusifolius*. By contrast it also includes a number of species that are amongst the rarest or most local in these islands, e.g. *Koenigia islandica*, *Polygonum maritimum*, *Rumex aquaticus* and *R. rupestris*.

Species of Polygonaceae occupy a range of habitats, from seashores – *Polygonum maritimum*, *P. oxyspermum* subsp. *raii*, *Rumex crispus* subsp. *littoreus* and *R. rupestris* – to mountain rocks and grassland, e.g. *Koenigia islandica*, *Oxyria digyna* and *Persicaria vivipara*, and from woodland rides and borders, e.g. *Fallopia dumetorum* and *Rumex sanguineus*, to marshes and the margins of streams, rivers, ponds and lakes. Species of *Persicaria* are often found in damp or wet habitats, for example *P. hydropiper*, *P. dubia* and *P. minor*, together with *Rumex aquaticus*, *R. hydrolapathum*, *R. palustris* and *R. maritimus*. *Persicaria amphibia* frequently grows in standing or gently flowing water, with the leaves floating on the surface, but a terrestrial form also occurs. *Rumex crispus* subsp. *uliginosus* is restricted to the muddy banks of tidal rivers. Some species occur in dry grassland, e.g. *Rumex pulcher* and *R. acetosella*. *R. acetosa* is less demanding than most other species in habitat preference and is found on many different types of soil at a variety of altitudes.

A number of published autecological accounts are available, mostly in parts of the Biological Flora of the British Isles published in *Journal of Ecology*. These are of *Fallopia japonica* (Beerling, Bailey & Conolly 1994), *Persicaria maculosa* and *P. lapathifolia* (Simmonds 1945, as *Polygonum persicaria* and *P. lapathifolium*), *P. hydropiper* (Timson 1966, as *Polygonum hydropiper*), *P. amphibia* (Partridge 2001), and *Rumex obtusifolius* and *R. crispus* (Cavers & Harper 1964). Cavers & Harper (1967a, 1967b) published further ecological data on *Rumex crispus*, especially plants in seashore habitats. There are also published ecological data on *Koenigia islandica* (Ratcliffe 1959) and *Rumex aquaticus* (Idle 1968). Useful accounts, based on non-British material, exist for *Polygonum oxyspermum* (Nordhagen 1963), *P. maritimum* (Beeftink, 1964), *Persicaria vivipara* (Engell 1973, 1978) and *Rumex acetosella* (Harris 1970, den Nijs 1984).

Many insects and their larvae feed on Polygonaceae, especially *Rumex* species, and after severe insect attack the leaves may be riddled with holes or even reduced to a filigree of veins. *Gastrophysa viridula*, a metallic green crysomelid leaf beetle, is perhaps the most important herbivorous invertebrate that feeds on *Rumex*, especially *R, crispus* and *R, obtusifolius*, and *R. hydrolap-athum* has its own beetle fauna. Several weavils, especially *Apion* spp, feed on *Rumex* species, including *R. acetosa* and *R. acetosella*. Salt & Whittaker (1996) provide an introduction and keys to these and other insect herbivores on docks and sorrels.

Rumex acetosella and *R. acetosa* are the food-plants of the larva of the Small Copper butterfly, *Lycaena phlaeas* L., and several species of Noctuid moths, and *R. hydrolapathum* is the food-plant of the larva of the Large Copper butterfly, *L. dispar* Howarth.

POLLINATION

The species of Polygonaceae exhibit a number of pollination mechanisms. *Persicaria bistorta* and *Fallopia japonica*, for example, have conspicuous inflorescences and well-developed perianths, with nectar at the base of the stamens, and are pollinated by insects. *Polygonum aviculare* and related species, on the other hand, have small flowers that are mostly odourless and without nectar and consequently are rarely visited by insects. These species self-pollinate with great efficiency. There is evidence, however, that ants pollinate some species of *Polygonum* (Hickman 1974, Akeroyd 1986).

Persicaria maculosa and related species have smaller flowers and less showy inflorescences than *P. bistorta* and they also lack perfume, but they do produce a small supply of nectar. Insect visits to the flowers are fairly frequent, making cross-pollination possible, although self-pollination often occurs.

In *Fagopyrum* the flowers are dimorphic and heterostylous, and nectar is contained in glands at the base of the eight stamens: the three inner dehisce extrorsely and the five outer introrsely, so that insect visitors are dusted with pollen. *Fagopyrum* is popular with bee-keepers, who plant it for its nectar.

Rumex has bulky, lax inflorescences, well exposed to the wind, small perianths, anthers pendent on long filaments, and large, brush-like stigmas. The flowers are pendulous on slender stalks and are wind-pollinated, although sometimes visited by bumble bees for pollen.

Rheum, like *Persicaria* and *Polygonum*, has flowers with capitate stigmas and nectaries close to the base of the stamens. It is pollinated by insects.

SEED DISPERSAL AND GERMINATION

The seeds of certain species, e.g. *Polygonum aviculare, Persicaria maculosa* and *P. lapathifolia*, are known to pass through the alimentary tracts of birds and animals without being damaged, and they are also distributed by adhering to clothes, footwear, fur and feathers when wet. The riparian species *Persicaria hydropiper*, *P. dubia* and *P. minor*, on the other hand, have seeds that are buoyant and are usually dispersed by water, as are those of *Polygonum maritimum and P. oxyspermum*. *Rumex* seeds, however, are much lighter and the fruits are often wind-dispersed over considerable distances. The corky tubercles on the fruits of many *Rumex* species probably aid dispersal by water, especially in *R. crispus* subsp. *littoreus* (Cavers & Harper 1967b), *R. crispus* subsp. *uliginosus* and *R. rupestris*. Flotation experiments indicate that fruits of *R. conglomeratus*, *R. crispus*, *R. hydrolapathum*, *R. maritimus* and *R. palustris* remain buoyant and viable for considerable periods (Lousley 1944). Several native and introduced *Rumex* species have fruits with hooked teeth or spines, which readily adhere to clothing, wool, fur and feathers. The fruits of *Emex* and *Oxygonum* are also dispersed in this way.

In common with the seeds of many weeds, some of those of *Polygonum*, *Persicaria* and *Rumex* can remain viable for long periods. Toole & Brown (1946) recorded 6% germination of seeds of *Rumex crispus* in a sample buried in soil for 40 years and Darlington & Steinbauer (1961) recorded 8% in a sample buried for 70 years. This can represent a significant number of survivors in that a single plant of *R. crispus* may produce up to 40,000 seeds (Cavers & Harper 1964). Seeds of *R. crispus*, *R. obtusifolius* and probably other *Rumex* species germinate in response to light, for example when soil is tilled or disturbed. Size and potential dormancy of seeds in *R. crispus* is dependent upon their position on the inflorescence, and seed dormancy varies considerably between individuals and populations (Lousley 1944, Cavers & Harper 1964, 1967b; Le Deunff 1974).

ECONOMIC USES

Polygonaceae contains relatively few species of economic importance, although several are noxious weeds of cultivation. *Rumex crispus* and *R. obtusifolius* are both proscribed weeds under the Weed Control Act 1958, whereby a landowner is obliged to destroy them to prevent infestation of adjacent property. This Act is one of the least frequently applied of any existing laws, but it does reflect the historical success of these species as weeds of cultivation. Several annual species of *Persicaria*, *Polygonum aviculare* and related species, and *Fallopia convolvulus* are frequent weeds of arable land and gardens. The worst weed of all is *Fallopia japonica*, which, together with *F. sachalinesis* and their hybrid *F. × bohemica*, forms thickets over large areas of wayside and derelict ground in the inner cities and suburbs, notably in London and around Swansea, and by streams and rivers in town and countryside. It is now an offence under Schedule 9 of the Wildlife and Countryside Act 1981 (amended 1985 and 1991) to plant these invasive taxa in the wild in Britain.

Rhubarb (*Rheum × officinale* and related hybrids) is widely grown, especially in the West Riding of Yorkshire and parts of the United States, for its fleshy, edible petioles. Foust & Marshall (1991) give a useful summary of the history of Rhubarb as a commercial food crop. Rhubarb root, which has purgative properties, was regarded as an important medicine during the 18th and 19th centuries and provided the basis of a considerable trade between Asia and Europe (Foust 1992). Chinese Rhubarb, derived from the dried rootstock of *Rheum officinale* and *R. palmatum* or their hybrid, remains in use as a gentle herbal laxative. The dried rhizome of *Fallopia japonica* has long been used in traditional Chinese and Japanese medicine to treat skin infections (Kimura *et al.* 1983), also arthritis, jaundice, bronchitis, physical injuries and burns, and in other countries as a treatment for skin infections and gout (Lewis & Elvin-Lewis 2003). *Polygonum aviculare* is employed as a herbal remedy in the alleviation of internal bleeding, and is a mild laxative. Yellow Dock (derived from *Rumex crispus*) is a herbal remedy used to treat a number of skin, respiratory and alimentary conditions.

The seeds of several species of Polygonaceae have been used at one time or another for human or animal food. The stomach contents of Tollund Man and other Iron Age corpses excavated from Danish peat-bogs revealed that the last meal of these unfortunates was a gruel of various seeds, including those of *Persicaria maculosa* and a *Rumex* species (Glob 1969). Although these bog burials are known to have been human sacrifices, perhaps involving a ritual meal, plenty of circumstantial archaeological evidence points towards the use of such seeds more generally in human food.

Fagopyrum esculentum is cultivated as a grain crop (Buckwheat) in several regions of the world, for example eastern Europe, especially on poorer soils and at high altitudes. In the United States and elsewhere it is a popular plant with bee-keepers, as it provides abundant nectar for a much-prized honey. In Britain it is grown as a cover- and food-plant for game birds and as a green manure in organic gardening. Its culinary popularity is increasing in Britain, notably as cooked groats or *kasha* ('porridge'), a staple of Russian, Ukrainian and Polish cuisine, or in the form of Breton pancakes or galettes made from the flour. Akeroyd (1993b) gives an account of this and other edible Polygonaceae. *F. esculentum* contains high levels of the flavonol glycocide rutin, used for the relief of high blood pressure and circulatory ailments (Spencer 1987, Lewis & Elwin-Lewis 2003).

Several species of Polygonaceae are or have been cultivated for ornament in gardens. Indeed, many of the species present in Great Britain and Ireland as naturalised aliens have their origin in gardens. *Persicaria wallichii*, *P. campanulata*, *Fallopia japonica* and *F. sachalinensis* were widely planted in the past. *Persicaria affinis*, *P. amplexicaulis* and *P. bistorta*, the latter usually as the robust cultivar 'Superba', are still popular flowers of the herbaceous border. *Persicaria vivipara* and *Oxyria digyna* are sometimes grown in rock-gardens or collections of alpines. *Rumex sanguineus* var. *sanguineus* has been grown since the Middle Ages as a medicinal herb and for its ornamental leaves. *Rheum palmatum* is a stately plant for the damp garden, and *Persicaria amphibia* f. *amphibia* and *Rumex hydrolapathum* are handsome additions to the larger water-garden.

Rumex acetosa and *R. patientia* have a long history of cultivation as potherbs, mostly in central and eastern Europe. *Rumex scutatus* and *R. rugosus* retain a niche in the contemporary garden for their use as flavouring, especially in salads and sorrel soup. *R. pseudoalpinus* has a history as a potherb and as a medicinal substitute for Rhubarb. *Persicaria bistorta* is an essential ingredient of Easter-ledges Pudding, a traditional herb pudding still eaten in the Lake District and Pennines, especially in Calderdale, Yorkshire, during Passiontide, the last two weeks of Lent (Grigson 1955, Mabey 1996). Another edible member of the family, Vietnamese Coriander (*Persicaria odorata* (Lour.) Soják), is sometimes grown in gardens as a culinary herb.

IDENTIFICATION OF POLYGONACEAE

Although many members of Polygonaceae can be identified from the seedling stage onwards, it is chiefly the characters of the ripe fruits that are used for critical determination, especially in *Rumex*. Immature material is, therefore, usually best left ungathered.

Since many species are common or are of alien origin, it is generally advisable, in cases where any doubt exists as to identity, to collect a voucher specimen. Hybrids especially are best checked against reference material. The herbarium of the University of Reading (**RNG**), especially, holds extensive collections of *Rumex*, *Persicaria* and *Polygonum*, including those assembled by the late Ted Lousley and Ted Wallace. The herbarium of the University of Cambridge (**CGE**) also has a considerable collection of *Polygonum*.

In order to identify *Polygonum* and *Persicaria* it is advisable, where possible, to collect plants complete with both lower and upper leaves, flowers and mature fruits. Hybrids are rare (Timson 1965) and are usually more or less intermediate between the putative parents (Danser 1921). Note that specimens of *Polygonum* collected from late September onwards often have anomalous nuts that are greenish and extending well beyond the perianth. In perennial species of *Fallopia* leaf characters are especially important, but it is advisable to collect flowering shoots where possible.

In *Rumex* the material required for determination is part of the main stem bearing the lowest obtainable leaves and a panicle bearing fully matured fruits. At the time of fruiting, however, the lower leaves of many species of *Rumex* have rotted away, and it is often tempting to gather leaves from any adjacent younger plants. This practice is to be discouraged as it is liable to lead to mixed gatherings and consequent confusion. Hybrids may also be present.

RECOGNISING *RUMEX* HYBRIDS

Hybrids are frequent in *Rumex* subgenus *Rumex* but have never been recorded between species in different subgenera. Indeed, hybrids occur at variable frequencies almost wherever different species of subgenus *Rumex* coexist. They are often intermediate between the putative parents and generally exhibit partial or almost complete sterility.

Non-hybrid plants pass regularly out of flower from the base to the summit of the panicle, so that when the lower whorls are passing into fruit those at the top of the panicle are still in flower. Eventually all whorls set fruit. In hybrids most flowers fail to set fruit (or produce empty fruits) and may dry off and fall without any appreciable enlargement of the three inner perianth-segments. A few here and there may be fertile and set fruit, the inner perianth-segments enlarging to varying degrees. Thus the panicle of a *Rumex* hybrid acquires a ragged, irregular appearance, and in most of the whorls will be found (1) stumps of the peduncles where flowers have dropped off, (2) shrivelled remains of perianths which have failed to develop, (3) partially enlarged perianth-segments and, occasionally, (4) matured valves containing empty or apparently good fruit.

Such hybrids are often conspicuously flushed with red. They may also display continuous and indeterminate growth, with new flowering shoots appearing at the end of the flowering season. These traits are particularly evident in the widespread hybrid between *R. crispus* and *R. obtusifolius* (*R.* × *pratensis*), which may form a significant proportion of mixed populations of the parent species. Backcrosses between the hybrid and its parents are harder to detect, although hybrid indices, as used by Williams (1971), and scatter diagrams of leaf characters, as used in Germany by Ziburski *et al.* (1986), are useful tools to detect intermediates. Similarly, putative backcrosses derived from *R. conglomeratus* and *R. sanguineus* can be recognised by careful recording of vegetative and fruit characters. Few studies of pollen fertility in hybrids have been undertaken apart from the work of Ziburski *et al.* (1986).

GENERAL SYNOPSIS AND KEYS TO POLYGONACEAE

Herbs, sometimes shrubs or climbers. Leaves usually alternate. Stipules united into a membranous sheath (ochrea). Flowers hermaphrodite or unisexual; perianth 3- to 6-merous, petaloid or green, herbaceous, the segments often enlarging and becoming membranous or corky in fruit. Stamens usually 6–9. Ovary superior, unilocular; styles 2–4; ovule solitary, basal. Fruit a trigonous, triquetrous or lenticular nut.

Key to genera of Polygonaceae

1 Perianth-segments 3 or 4 (mountain plants)   2
  2 Perianth-segments 3; annual   2 **Koenigia** (p. 27)
  2 Perianth-segments 4; perennial   10 **Oxyria** (p. 34)
1 Perianth-segments 5 or 6   3
  3 Perianth-segments 6   4
    4 Outer perianth-segments armed with 3 stout terminal spines (rare
      casual)   11 **Emex** (p. 35)
    4 Outer perianth-segments without stout spines, but margins often with
      flexible spines   5
      5 Leaves palmately lobed or entire; stamens 9   8 **Rheum** (p. 30)
      5 Leaves never palmately lobed; stamens 6   9 **Rumex** (p. 30)
  3 Perianth-segments 5 (or 4–5 in *Oxygonum*)   6
    6 Outer perianth-segments armed with 3 stout spines (rare casual)
                                                    4 **Oxygonum** (p. 27)
    6 Outer perianth-segments without spines   7
      7 Perianth succulent in fruit   8
        8 Perianth purplish-black in fruit; erect perennial herb
                  1 (Species 4) **Persicaria mollis** (p. 23)
        8 Perianth waxy-white in fruit; woody climber or scrambler
                                      7 **Muehlenbeckia** (p. 30)
      7 Perianth sometimes enlarged, but not succulent in fruit   9
        9 Outer perianth-segments keeled or winged in fruit
                                           6 **Fallopia** (p. 29)
        9 Outer perianth-segments not keeled or winged in fruit   10

10 Nut at least twice as long as perianth; leaves deltate, about
    as wide as long   3 **Fagopyrum** (p. 27)
10 Nut usually less than twice as long as perianth; leaves not
    deltate, at least twice as long as wide (and usually much
    longer)   11
    11 Flowers in terminal or axillary racemes, panicles or
        spike-like inflorescences; ochreae truncate, usually
        ciliate or fimbriate   1 **Persicaria** (p. 23)
    11 Flowers solitary or in small axillary clusters; ochreae
        lacerate, silvery or hyaline   5 **Polygonum** (p. 28)

Synopses of genera and keys to their subgenera, sections and species

1 **PERSICARIA** (L.) Mill.

Annual or perennial herbs, sometimes subshrubby. Leaves variously shaped,
longer than wide, pinnately veined. Ochreae usually truncate, ciliate or fimbriate. Flowers usually bisexual, in cymes arranged in terminal or axillary panicles,
racemes or spike-like or capitate inflorescences. Perianth-segments usually 5
but sometimes 4, ± equal, petaloid, not winged or keeled. Stamens 5–8. Styles
2, sometimes 3. Nut trigonous, triquetrous or lenticular, enclosed in the persistent perianth or protruding from it for less than half its length.

1 Leaves mostly basal; stem simple; flowers in 1–2(–more) short, dense
   racemes   Section 2 *Bistorta* (p. 25)
1 Leaves mostly cauline; stem often branched; flowers in panicles, racemes
   or spike-like or capitate inflorescences   2
   2 Perennial or subshrubby; flowers in diffuse terminal panicles   3
      3 Ochreae fringed with bristles or stiff hairs; shoots hairy, eglandular
                                         Section 1 *Aconogonon* (p.24)
      3 Ochreae entire; shoots subglabrous, usually glandular
                                         Section 4 *Amblygonon* (p. 25)
   2 Annual or rarely perennial; flowers in panicles, racemes or spike-like or
      capitate inflorescences   4
      4 Styles 2, rarely 3; flowers in spike-like inflorescences
                                         Section 5 *Persicaria* (p. 26)
      4 Styles 3; inflorescences ± capitate   5
         5 Stems with deflexed prickles, glabrous or pubescent
                                         Section 3 *Echinocaulon* (p. 25)
         5 Stems without prickles, glabrous   Section 6 *Cephalophilon* (p. 27)

Section 1: *ACONOGONON* (Meisn.) H. Gross

Robust or subshrubby rhizomatous perennials without basal leaf-rosettes. Stems stout, ± branched, glabrous or somewhat hairy. Leaves often large, variable in shape and hairiness. Ochreae ± tubular, truncate or pointed. Flowers in branched terminal panicles, basically bisexual but in some species functionally unisexual. Perianth of 5 ± unequal petaloid segments. Stamens 8. Styles 3. Nut trigonous. Considered by some botanists to be a separate genus, *Aconogonon* (Meisn.) Reichenb. (but see account of 2 *P. wallichii*).

1 Perianth pink or red   3 **P. campanulata**
1 Perianth white, greenish-white or cream   2
  2 Most leaves more than 4 times as long as wide; inflorescence diffuse
        1 **P. alpina**
  2 Most leaves about 3 times as long as wide; inflorescence leafy   3
    3 Panicles lax; perianth-segments unequal   2 **P. wallichii**
    3 Panicles dense; perianth-segments ± equal   4 **P. mollis**
  2 Most leaves less than 3 times as long as wide; inflorescence pyramidal, fairly dense   4
    4 Underside of leaves densely white-tomentose   5 **P. weyrichii**
    4 Underside of leaves usually slightly pubescent with some appressed hairs   sub 1 **P. × fennica**

Vegetative key (after P.D. Sell)

1 Leaves glabrous or with few hairs on upper surface   2
  2 Lower leaves truncate to cordate at base   2 **P. wallichii** var. **wallichii**
  2 Lower leaves cuneate at base   3
    3 Upper leaves ensiform to narrowly lanceolate   1 **P. alpina**
    3 Upper leaves broadly lanceolate   sub 1 **P. × fennica**
1 Leaves with numerous hairs on upper surface   4
  4 Leaves appressed-hairy on upper surface   4 **P. mollis**
  4 Leaves hairy on upper surface, but hairs not appressed   5
    5 Underside of leaves with numerous short hairs, not concealing surface
        2 **P. wallichii** var. **pubescens**
    5 Underside of leaves with dense hairs, ± concealing surface   6
      6 At least some leaves more than 8 cm wide, densely white-tomentose beneath   5 **P. weyrichii**
      6 Leaves not more than 8 cm wide, with whitish to buff pubescence beneath   3 **P. campanulata**

Section 2: *BISTORTA* (L.) Jaretzky (*Polygonum* section *Bistorta* (L.) D. Don)

Perennial herbs with basal leaf-rosettes. Stems branched or (usually) unbranched, erect, from a rhizomatous or woody rootstock. Leaves ovate to linear-lanceolate, petiolate. Ochreae truncate. Flowers in terminal spike-like racemes, bisexual. Perianth of 5 subequal petaloid segments. Stamens 8 (sometimes unequal, often exserted). Styles 3. Nut trigonous, enclosed in the perianth. Considered by some botanists to be a separate genus, *Bistorta* (L.) Adans.

1 Lamina of basal leaves tapering at the base; lower part of inflorescence replaced by bulbils   6 **P. vivipara**
1 Lamina of basal leaves truncate, cordate or subcordate at the base; inflorescence without bulbils   2
  2 Petioles of lower leaves winged; upper leaves not amplexicaul
                                                     7 **P. bistorta**
  2 Petioles not winged; upper leaves amplexicaul   8 **P. amplexicaulis**

Section 3: *ECHINOCAULON* Meisn.

Annuals. Stems weak, with minute recurved prickles. Leaves ovate-lanceolate to ovate or oblong. Ochreae truncate, membranous, slightly pilose. Flowers in short subcapitate racemes. Perianth of 4 or 5 petaloid segments. Styles 3. Nut triquetrous. (Rare aliens.)

Sections 3 and 6 are considered by some botanists to comprise a separate genus, *Truellum* Houtt.

1 Leaves not more than 6 cm long, sagittate at the base; perianth-segments 4 (Co. Kerry, probably extinct)   9 **P. sagittata**
1 Leaves usually at least 5 cm long, cuneate at the base; perianth-segments 5
                                                   10 **P. bungeana**

Section 4: *AMBLYGONON* Meisn.

Similar to section *Persicaria* and included here in the key to that section (see 5 below), but robust rhizomatous perennials. Ochreae without bristles or cilia. Inflorescence a branched panicle of elongate spike-like racemes. Nut lenticular. (Rare alien.)   11 **P. senegalensis**

Section 5: *PERSICARIA* (L.) H. Gross

Usually annuals, without basal leaf-rosettes. Stems branched, erect or decumbent. Leaves lanceolate to ovate-lanceolate, petiolate. Ochreae truncate, ± membranous, brownish, often with a fringe of bristles or cilia. Flowers in dense or lax terminal and axillary spike-like inflorescences, usually bisexual. Perianth of 3–5 ± petaloid segments. Stamens 4–8. Styles 2–3. Nut trigonous or lenticular, enclosed in the perianth.

1 Rhizomatous perennial, usually aquatic and often floating; stamens
   sometimes exserted   12 **P. amphibia**
1 Annual (rarely perennial), terrestrial but often riparian; stamens included 2
  2 Spike-like inflorescences dense, with the flowers crowded, overlapping 3
    3 Peduncles eglandular (rarely with a few subsessile yellow glandular
       hairs or sessile glands)   4
      4 Leaves with apex sharply acuminate and margins bearing a row of
         short yellow spines (rare alien)   15 **P. glabra**
      4 Leaves with apex acute (but not acuminate) and margins without
         spines   13 **P. maculosa**
    3 Peduncles with glandular hairs or sessile glands   5
      5 Plant very robust, up to 200 cm or more tall; leaves more than
         15 cm long (rare alien)   11 **P. senegalensis**
      5 Plant rarely more than 100 cm tall; leaves usually less than 15 cm
         long   6
        6 Glands on peduncles distinctly stipitate; perianth bright pink to
           purplish-pink   16 **P. pensylvanica**
        6 Glands on peduncles subsessile, yellow; perianth dull pink or
           greenish-white   14 **P. lapathifolia**
  2 Spike-like inflorescences lax and slender, with each flower distinctly
    visible   7
    7 Perianth greenish or reddish, with flat brownish or yellowish glands;
       nut dull   17 **P. hydropiper**
    7 Perianth purplish-pink, with small colourless glands, or glands absent;
       nut glossy   8
      8 Leaves narrowly elliptical, usually less than 5 times as long as
         wide; nut usually at least 3 mm long   18 **P. dubia**
      8 Leaves linear-lanceolate to linear-oblong, usually more than
         5 times as long as wide; nut not more than 2.5 mm long
                                       19 **P. minor**

Section 6: *CEPHALOPHILON* Meisn.

Annual or perennial herbs. Stems branched, creeping or straggling. Leaves oblong to ovate, acute to acuminate. Ochreae obliquely truncate, with stiff white hairs, deciduous. Flowers in solitary, capitate globose heads. Perianth of 4 or 5 petaloid segments. Styles 3. Nut lenticular, enclosed in the persistent perianth.

Sections 3 and 6 are considered by some botanists to comprise a separate genus, *Truellum* Houtt.

1 Straggling annual; inflorescence subtended by a leafy bract;
   perianth-segments 4   20 **P. nepalensis**
1 Creeping or trailing perennial; inflorescence not subtended by a leafy bract;
   perianth-segments 5   21 **P. capitata**

2 KOENIGIA L.

Diminutive glabrous annual herb. Leaves mainly subopposite, elliptical. Ochreae brown, hyaline. Flowers bisexual, subsessile or with short petioles, in terminal and axillary cymose clusters. Perianth-segments 3, sepaloid. Stamens usually 3, alternating with 3 gland-like staminodes. Styles 2(–3), very short. Nut bluntly trigonous, ± enclosed by the perianth. (N.W. Scotland.)

                                                          22 **K. islandica**

3 FAGOPYRUM Mill.

Annual or perennial herbs. Stem usually single, erect, hollow, branched. Leaves sagittate or hastate, palmately veined, petiolate. Ochreae membranous, not ciliate. Flowers bisexual, heterostylous, in axillary and terminal umbels. Perianth-segments 5, subequal, not enlarged in fruit. Stamens 8 (5 outer and 3 inner). Styles 3, long and slender. Nut sharply trigonous, much exceeding the perianth.

1 Plant perennial, up to 1 m or more tall (rare alien)   25 **F. dibotrys**
1 Plant annual, frequently not more than 50 cm tall   2
   2 Perianth-segments 3–4 mm long; nut smooth   23 **F. esculentum**
   2 Perianth-segments *c*. 2 mm long; nut rugose (rare alien) 24 **F. tataricum**

4 OXYGONUM Burch. ex Campd.

Annual or perennial herbs. Leaves entire or lobed. Flowers polygamous, in axillary spikes. Perianth-segments 5 (4–5 in male flowers). Stamens 8. Styles 3. Perianth enclosing the fusiform nut, with 3 stout, spreading spines *c*. 2 mm long arising near the middle. (Rare casual.)   26 **O. sinuatum**

5 POLYGONUM L. (*Polygonum* section *Avicularia* Meisn.)

Annual, biennial or perennial herbs. Stems branched, mostly more or less prostrate or only weakly erect. Leaves small, narrowed at the base, much longer than
wide, pinnately veined. Ochreae bipartite or lacerate, more or less silvery or
membranous. Flowers small, solitary or few, subsessile or on short pedicels, in
axils of leaves. Perianth-segments 5, equal, often greenish but ± petaloid.
Stamens 5–8. Styles (2–)3. Nut trigonous or lenticular, enclosed in the persistent perianth or protruding from it for less than half its length.

Members of this genus are easily recognised by their low habit with diffuse
stems, small and often white or greenish flowers in the axils of the leaves, and
white, silvery or hyaline ochreae.

1 Perennial with woody rootstock; leaves usually greyish   2
   2 Upper leaves (bracts) much smaller, often scarious; nut not more than
     2.5 mm long (rare casual)   **34 P. equisetiforme**
   2 All leaves subequal in size; nut usually at least 3 mm long   3
     3 Nut more than 2.5 mm wide, distinctly exceeding the perianth;
     ochreae shorter than the internodes (seashores)
         29 P. oxyspermum subsp. **raii**
     3 Nut not more than 2.5 mm wide, equalling or slightly exceeding
     the perianth; ochreae often longer than the internodes, at least in
     inflorescence   4
       4 Leaves greyish; nut 4–4.5 mm long (seashores)   **28 P. maritimum**
       4 Leaves green; nut *c*. 3 mm long (rare casual)   **33 P. cognatum**
1 Annual; leaves usually green   5
   5 All leaves equal or subequal in size   6
     6 Leaves greyish or green; nut 3–6 mm long, distinctly exceeding the
     perianth (seashores)   **29 P. oxyspermum** subsp. **raii**
     6 Leaves green; nut less than 3 mm long, enclosed within or slightly
     exceeding the perianth   7
       7 Perianth-segments united for up to half their length; nut brown,
       dull; stems glabrous   **33 P. arenastrum**
       7 Perianth-segments united for less than half their length; nut dark
       brown or black, shining; stems minutely puberulent (rare casual)
         35 P. plebejum
   5 Leaves of lateral branches distinctly smaller than those of main stem   8
     8 Leaves of main stem usually caducous, the uppermost (bracts) much
     smaller, often scarious; nut 1.5–3 mm long (rare casuals)   9

9 Perianth-segments green with purplish or pinkish margins; nut
  2.5–4 mm long, glossy   36 **P. bellardii**
9 Perianth-segments pink or white; nut *c.* 2 mm long, dull
                                          37 **P. arenarium** subsp. **pulchellum**
  8 Leaves of main stem not caducous (sometimes deciduous by autumn)
    or scarious, subequal; nut 2.5–4.5 mm long   10
    10 Leaves usually less than 5 mm wide; perianth-segments not
       overlapping, often reddish   31 **P. rurivagum**
    10 Leaves usually more than 5 mm wide; perianth-segments
       overlapping, white to pink or greenish   11
       11 Leaves of main stem subsessile or with petiole usually no
          more than 2 mm long; nut 2.5–3.5 mm long   30 **P. aviculare**
       11 Leaves of main stem with petiole 4–8 mm; nut 3.5–4.5 mm
          long (Scotland)   32 **P. boreale**

6 FALLOPIA Adans. (*Bilderdykia* Dumort.; incl. *Reynoutria* Houtt.)

Annual herbs, robust perennial herbs with extensive rhizomes, forming large
clumps, or perennial climbing shrubs. Leaves ovate to oblong, cordate or sagittate at the base, petiolate. Flowers unisexual or bisexual (plants dioecious:
female flowers with substantial rudiments of stamens, the male with non-func-
tional ovaries and stigmas), in axillary panicles. Perianth-segments 5(–6), the
outer 3 larger, keeled or winged, often enlarging in fruit. Stamens 8. Styles 3.
Nut trigonous, not exceeding the perianth.

1 Annual with twining stems   2
  2 Fruiting pedicels 1–3 mm; nut 4–5 mm long, dull   38 **F. convolvulus**
  2 Fruiting pedicels 5–8 mm; nut not more than 3 mm long, glossy
                                                      39 **F. dumetorum**
1 Perennial with erect or twining stems   3
  3 Vine-like, woody climber   40 **F. baldschuanica**
  3 Robust perennial with erect stems   4
    4 Leaves seldom more than 12 cm long, truncate at base, cuspidate;
      flowers white   41 **F. japonica**
    4 Leaves usually more than 15 cm long, cordate at base, usually acute;
      flowers greenish   5
      5 Leaves distinctly cordate at base, usually acute, with sparse
        long hairs beneath   42 **F. sachalinensis**
      5 Leaves weakly to moderately cordate at base, acuminate, with
        numerous short hairs beneath   43 **F. × bohemica**

7 MUEHLENBECKIA Meisn.

Scrambling shrubs (or perennial herbs). Leaves broadly oblong to suborbicular. Flowers unisexual (plants dioecious). Perianth deeply 5-partite, sepaloid, enlarging and fleshy in fruit. Stamens 8. Styles 3. Nut trigonous, not exceeding the perianth.   **44 M. complexa**

8 RHEUM L.

Robust perennial herbs with a woody rhizome. Leaves large, mainly basal, with palmate veins. Flowers bisexual, in a large panicle. Perianth-segments 6, free, equal, not enlarging in fruit. Stamens 6 or 9. Styles 3. Nut cordate-ovoid, strongly winged.

1 Leaves entire (deeply cordate at base)   **46 R. × hybridum**
1 Leaves toothed or palmately lobed   2
  2 Leaves deeply palmately lobed, with acute divisions, coarsely toothed
                                                   45 R. palmatum
  2 Leaves ovate, 5-lobed, with lobes deeply toothed   **47 R. officinale**

9 RUMEX L.

Annual, biennial or perennial herbs. Leaves alternate, variously shaped, longer than wide. Flowers bisexual or unisexual (plants polygamous or dioecious), in whorls in simple or branched panicles. Perianth-segments 6, the 3 outer small and thin and remaining virtually unaltered, the 3 inner (valves) usually enlarging greatly in fruit. Stamens 6. Styles 3. Nut trigonous, enclosed by the perianth.

Rumex can be divided into four distinct subgenera, recognised as genera by some botanists.

1 Plants male or female; basal and lower cauline leaves hastate or sagittate, acid to the taste   2
  2 Inner perianth-segments not enlarging in fruit
                                Subgenus 1 *Acetosella*
  2 Inner perianth-segments enlarging greatly in fruit
                                Subgenus 2 *Acetosa*
1 Plants hermaphrodite; basal and lower cauline leaves never hastate or sagittate, not acid to the taste   3
  3 Valves more than twice as wide as the nut   Subgenus 3 *Rumex*
  3 Valves less than twice as wide as the nut (rare alien)
                               Subgenus 4 *Platypodium*

Subgenus 1: *ACETOSELLA* Raf.

Creeping perennials. Leaves hastate or sagittate, tasting of acid. Flowers unisexual (plants dioecious). Inner perianth-segments not enlarging in fruit or at most little exceeding the nut, without tubercles. 2x = 14.   48 **R. acetosella**

Subgenus 2: *ACETOSA* (Mill.) Rech. fil.

Perennials. Leaves mostly hastate or sagittate, tasting of acid. Flowers unisexual (plants dioecious or polygamous). Inner perianth-segments enlarging in female flowers to several times the length of the nut, without tubercles. $2n = 2x = 14$ (female), 15 (male).

1 Plants hermaphrodite (polygamous); leaves not more than twice as long as wide (rare alien)   49 **R. scutatus**
1 Plants male or female; leaves 2–6 times as long as wide   2
  2 Panicle lax, rarely dense, with branches usually simple   50 **R. acetosa**
  2 Panicle dense, with branches repeatedly branched (rare alien)
                                                      sub 50 **R. rugosus**

Subgenus 3: *RUMEX*

Annuals, biennials or perennials. Basal leaves linear-lanceolate to oblong or ovate or sometimes panduriform, cuneate to rounded or cordate at the base, never hastate or sagittate, not tasting of acid. Stem leaves smaller and narrower. Flowers bisexual. Perianth-segments enlarging greatly in fruit to form papery valves much longer than the nut, each with or without a corky tubercle on the midvein. 2x = 20.

1 Valves entire or with a few small teeth, rarely crenulate   2
  2 Valves as wide as or wider than long   3
    3 Valves without tubercles   4
      4 Plant rhizomatous; basal leaves orbicular or suborbicular
        (N. England and Scotland)   53 **R. pseudoalpinus**
      4 Plant not rhizomatous; leaves lanceolate   5
        5 Leaves 3–4 times as long as wide, broadly lanceolate,
          not crisped; valves 5–6.5 mm wide   55 **R. longifolius**
        5 Leaves 8–5 times as long as wide, linear-lanceolate, crisped;
          valves less than 5 mm wide (rare alien) 56 **R. pseudonatronatus**
    3 1–3 of the valves with a tubercle   6

6 Stems with axillary flowering branches below, eventually
   overtopping primary panicle; leaves linear-lanceolate (rare alien)
                                                   51 **R. salicifolius**
6 Stems without axillary flowering branches overtopping primary
   panicle; leaves variously shaped but never linear-lanceolate   7
   7 Leaves strongly crisped   61 **R. crispus**
   7 Leaves uncrisped or scarcely crisped   8
      8 Leaves broadly cordate, scabrid-papillose beneath on veins
         (rare alien)   57 **R. confertus**
      8 Leaves not broadly cordate, sometimes scabrid beneath   9
         9 Plant not more than 40 cm tall; branches divaricate
            (rare casual)   73b **R. pulcher** subsp. **anadontus**
         9 Plant more than 40 cm tall, often more than 150 cm;
            branches ascending   10
            10 Usually 1 valve with a tubercle; if 3, then 1 tubercle
                considerably larger than the others   60 **R. patientia**
            10 All 3 valves with a well-developed tubercle
                                     61b **R. crispus** subsp. **uliginosus**
2 Valves longer than wide   11
11 Valves without tubercles (Loch Lomond)   54 **R. aquaticus**
11 1–3 of the valves with a tubercle   12
   12 Valves not more than 3 mm long   13
      13 Fruiting pedicel much longer than perianth; 1 valve with
          a large tubercle, the other 2 without or with small tubercles
                                                    65 **R. sanguineus**
      13 Fruiting pedicel about as long as or slightly longer than
          perianth; all 3 valves with large tubercles
                                                 64 **R. conglomeratus**
   12 Valves more than 3 mm long   14
      14 Plant rhizomatous, far-creeping; stem often with secondary
          axillary flowering branches below   52 **R. cuneifolius**
      14 Plant not rhizomatous; stem without secondary axillary
          flowering branches below   15
         15 Leaves glaucous (coasts of S.W. and W. Britain)
                                                    66 **R. rupestris**
         15 Leaves not glaucous   16
            16 Basal leaves lanceolate, more than 40 cm long;
                valves triangular   58 **R. hydrolapathum**
            16 Basal leaves ovate-oblong, not more than 40 cm
                long; valves lingulate or narrowly ovate-triangular
                (rare alien)   74c **R. obtusifolius** subsp. **sylvestris**

1 Valves with prominent teeth   17
18 Margin of valves with distinctly hooked teeth (rare casuals)   19
    19 Valves usually without tubercles
       20 Stem stout; whorls many-flowered; valves with 6–8 teeth on each margin   67 **R. steudelii**
       20 Stem slender, flexuous; whorls few-flowered; valves with 3–5 teeth on each margin   70 **R. brownii**
    19 1–3 of the valves with a tubercle   21
       21 Valves 3–4 mm, each margin with 5–6 teeth   68 **R. bequaertii**
       21 Valves 4–7 mm, each margin with more than 6 teeth   69 **R. nepalensis**
18 Margin of valves with straight or triangular teeth   22
    22 Valves lingulate (rare casuals)   23
       23 Young stems rough-papillose   79 **R. fueginus**
       23 Young stems smooth   24
          24 Valves not more than 2.5 mm; whorls confluent towards apex   71 **R. crystallinus**
          24 Valves 2.5–3 mm; flowers in remote whorls   72 **R. tenax**
    22 Valves rhombic or rhombic-triangular   25
       25 Valves 2–4 mm   26
          26 Mature plant brown or reddish-brown; marginal teeth stiff, as long as the width of the valve   77 **R. palustris**
          26 Mature plant golden-yellow; marginal teeth flexible, 2–3 times as long as the width of the valve   78 **R. maritimus**
       25 Valves 4–6 mm   27
          27 Only 1 valve with a tubercle   28
             28 Mature panicle with little fertile seed; marginal teeth mostly short, sometimes few   62 **R. × pratensis**
             28 Mature panicle fertile; marginal teeth up to half width of valves   74a **R. obtusifolius** subsp. **obtusifolius**
          27 All 3 valves with tubercles   29
             29 Tubercles of about equal size   30
                30 Leaves obovate; panicle leafy (rare casual)   76 **R. obovatus**
                30 Leaves not obovate; panicle leafless or with a few leafy bracts   31
                   31 Basal leaves panduriform; branches divaricate   73a **R. pulcher** subsp. **pulcher**
                   31 Basal leaves not panduriform; branches usually ascending (rare casuals)   32

32   Valves orbicular, with crenulate margins

63 **R. stenophyllus**

32   Valves deltate, with subulate marginal teeth

75 **R. dentatus**

29 One tubercle larger   33

      33   Valves 5–9 mm long, orbicular-ovate   59 **R. cristatus**

      33   Valves not more than 6 mm long, ovate-triangular   34

           34   Plant not more than 40 cm; leaf-veins glabrous
or puberulent beneath (rare casual)

73c **R. pulcher** subsp. **woodsii**

           34   Plant more than 40 cm; leaf-veins usually papillose-
scabrid beneath   35

              35   Mature panicle with little fertile seed, often
regenerating vigorously from base

62 **R.** × **pratensis**

              35   Mature panicle fertile, rarely regenerating from
base   74b **R. obtusifolius** subsp. **transiens**

Subgenus 4: *PLATYPODIUM* Willk.

Usually reddish annuals, with numerous short, slender stems. Leaves small, the basal ovate or spathulate, the stem leaves ovate or lanceolate, cuneate at the base. Flowers bisexual, in clusters of 4 or fewer. Flowers, fruits and pedicels often dimorphic. Perianth-segments very small in fruit, usually with teeth and small tubercles. $2n = 2x = 16$. (Rare casual.)   80 **R. bucephalophorus**

10 OXYRIA Hill

Perennial herbs. Leaves mostly basal, reniform. Flowers unisexual (plants dioecious). Perianth-segments 4, the outer 2 not accrescent, the inner 2 enlarging somewhat in fruit and becoming scarious and tinged with red, without tubercles. Stamens 6. Stigmas 2. Nut lenticular, broadly winged, much longer than the inner perianth-segments. (Mountains.)   81 **O. digyna**

11 EMEX Campd.

Annuals. Leaves ovate, truncate or cordate at the base. Flowers unisexual (plants monoecious), the female at the base of the inflorescence. Perianth-segments 6, free in male flowers, connate in female flowers, the outer 3 spinescent and indurate in fruit. Stamens 4 or 6. Styles 3. Nut trigonous, not exceeding the perianth. (Rare casuals.)

1 Fruiting perianth-segments less than 5 mm wide, the inner segments
   subacute   82 **E. spinosa**
1 Fruiting perianth-segments more than 5 mm wide, the inner segments more
   or less rounded, with a terminal spiny arista   83 **E. australis**

Note on distribution maps

In the species accounts that follow, distribution maps are included for most native and some alien species (except for some very common natives) and some subspecies. The maps show 10-km square records distinguished by two date classes: 1970 onwards (solid circles) and pre-1970 (open circles). Most maps show all records irrespective of the native or alien status of the species, which in this family of ruderals may not readily distinguished. However, the maps of *Polygonum maritimum*, *Fallopia dumetorum* and *Oxyria digyna*, species that have well-defined native ranges and outlying alien records, denote alien records with a separate symbol for 1970 onwards (×) and for pre-1970 records (+). Up-to-date information on distribution in Britain and Ireland is available on the BSBI website (www.bsbi.org.uk).

POLYGONACEAE A.L. Jussieu

1 **Persicaria alpina** (All.) H. Gross   *Alpine Knotweed*

Polygonum alpinum All.; *Aconogonon alpinum* (All.) Schur

Perennial with a short creeping underground rhizome and erect, branched, solid, smooth, often reddish stems 30–80(–130) cm. Leaves 8–15 × 1–3.5 cm, lanceolate to oblong-lanceolate, tapering at both ends, acute to acuminate, usually with short stiff hairs on both surfaces and on the margins; petiole up to 1 cm. Ochreae funnel-shaped, mainly hyaline, pale brown, with long hairs, soon disintegrating. Inflorescence a broad, pyramidal, diffuse, terminal panicle. Flowers bisexual, homostylous. Perianth-segments 5, 2–3.5 mm, subequal, white, pale yellowish-white or pinkish. Nut slightly exceeding the perianth, 3–5 mm, trigonous, pale brown, glossy. 2n = 20 (Jaretzky 1928). Flowering in July and August.

A native of montane regions from the Alps and the Balkans to Siberia, extending south into S.W. Asia; also in North America. Long grown in British and Irish gardens (first recorded in 1816) and an established escape in a few places in northern Britain. It has been recorded, for example, from a grassy slope on Cambuslang golf course, Lanarks (v.c. 77), where it has apparently become been established, and waste ground at Montrose and Ninewells, Dundee, Angus (v.c. 90). It has long been naturalized on a shingle island of the River Dee near Ballater, S. Aberdeen (v.c. 92).

The taxonomy of this and other species in section *Aconogonon* was discussed by Hong (1992).

Hybrid

Persicaria × *fennica* (Reiersen) Stace (*Aconogonon* × *fennicum* Reiersen) (1 *P. alpina* × 5 *P. weyrichii*)

Intermediate between the parents. Leaves narrowly ovate, broader than those of *P. alpina*, acuminate. Nut unwinged, with low fertility. Known since 1981 from New Mill, near Holmfirth, S.W. Yorks (v.c. 63), and identified by J. Reiersen as similar to plants widespread in Fennoscandia, especially Finland (Stace 2002). This hybrid is increasingly attracting attention in popular horticultural literature, and may naturalize more frequently in Britain in the future.

Persicaria alpina 1

Polygonum polystachyum Wall. ex Meisn.; *Persicaria polystachya* (Wall. ex Meisn.) H. Gross, non Opiz; *Aconogonon polystachyum* (Wall. ex Meisn.) M. Král.; *Pleuropteropyrum polystachyum* (Wall. ex Meisn.) Javeid & Munshi; *Rubrivena polystachya* (Wall. ex Meisn.) M. Král.

Perennial, with a creeping rhizome and stout, erect, solid, finely striate, reddish stems, glabrous except just below the nodes, (60–)80–200 cm, much branched above. Leaves 8–25 × 3–8 cm, lanceolate to oblong-lanceolate, acuminate-caudate to a long point, cordate or subcordate at the base, typically with minute, pale, eglandular hairs on the wavy margins and the often reddish veins beneath but ± glabrous above; petiole 0.5–3 cm. Ochreae of upper stems crowded, sometimes exceeding the internodes, pointed, entire, brown, typically glabrous, persistent. Inflorescence a lax, rather leafy, terminal or axillary panicle. Flowers bisexual, long- or short-styled. Perianth-segments 5, 2.8–3.5(–4) mm, unequal, the outer two much narrower than the three broader inner ones, white or pinkish. Nut included in the perianth, 3 x 2 mm, ovoid, trigonous with blunt angles, acute, smooth. $2n = 22$ (Jaretzky 1928). Flowering from August to October.

A native of the Himalayas and W. China, naturalised in W. and C. Europe. It has been grown in gardens in Britain and Ireland since the late 19th century. Though seed is only occasionally set here, *Persicaria wallichii* is thoroughly established as an escape by vegetative spread from outcast rhizomes, forming dense stands on roadsides, waste ground, railway embankments and the banks of streams and rivers. First recorded in the wild in 1917, it is now widely distributed. It has been particularly impressive for over 70 years in N. Devon (v.c. 4), especially along the disused Barnstaple–Lynton railway line, about Woody Bay, and by the River Heddon near Parracombe. It is now locally common in S.W. England, Wales and Ireland. Conolly (1977) has documented the history of this species in Great Britain and Ireland.

Hong (1993) and Karlsson (2000) place this species in *Persicaria* sensu stricto on cladistic and chromosomal grounds.

This is a variable species. *Persicaria wallichii* var. *pubescens* (Meisn.) Akeroyd, with numerous, long, pale, eglandular hairs on both surfaces and the margins of the leaves and on the ochreae, occurs frequently in Britain, especially in West Wales. It has sometimes been confused with 4 *Persicaria mollis* (from which it is best distinguished by its unequal perianth segments) or mistakenly referred to another cultivated species, *Polygonum lichiangense* (Conolly 1991; see under 3 *Persicaria campanulata*).

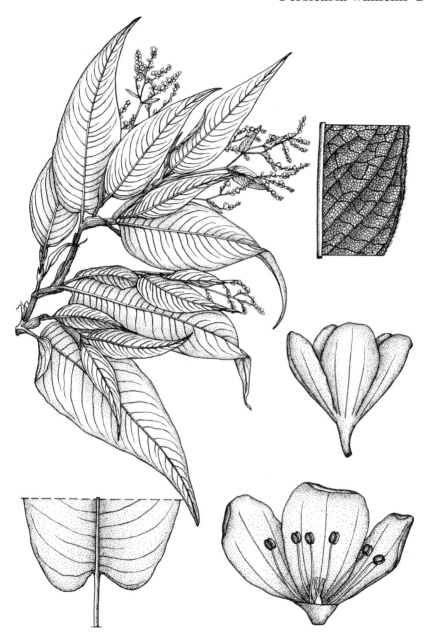

3 Persicaria campanulata (Hook. fil.) Ronse Decr. *Lesser Knotweed*

Polygonum campanulatum Hook. fil.; *Aconogonon campanulatum* (Hook. fil.) H. Hara

A handsome perennial with a creeping rhizome and hollow, strongly ridged, pubescent or tomentose stems up to 150 cm, forking widely above. Leaves 5–15 × (3–)5–8 cm, variable in shape and size, lanceolate to ovate or ovate-elliptic, acuminate at the apex, cuneate or rounded at the base, dark green above, prominent in herring-bone pattern, with or without a blackish blotch along the midrib, with numerous short to medium, stiff, appressed hairs above and with a fine tomentum of whitish to buff tangled threads beneath, with short, stiff hairs on the veins and margins; petiole up to 3 cm. Ochreae loosely tubular, membranous, conspicuous. Inflorescence a short terminal panicle with 3- to 5-flowered cymes. Flowers bisexual, heterostylous, scented. Perianth-segments 5, 4–5 mm, bright pink, rarely red. Stamens 8, with long filaments 2–3 mm. Nut not exceeding the perianth, 2 mm, ellipsoid, triquetrous, brown, glossy. $2n = c.$ 64, 60 (Kurosawa 1971). Flowering from July to September.

A native of the Himalayas and W. China, introduced to Britain and Ireland as a garden plant by 1909. First reported in the wild in 1933, it soon became established by streams and on marshy ground in areas of original planting. Though seed set has not been reported, it is truly naturalised in many places, especially in W. Scotland, N. and S.W. Ireland and the West Country, locally forming extensive patches on roadsides, streamsides and former railway tracks.

Conolly (1977) has documented the history of this species in Great Britain and Ireland. The taxonomy of this and other species in section *Aconogonon* was discussed by Hong (1992).

Polygonum lichiangense W.W. Smith has been recorded from Pembroke (v.c. 45) in error for 2 *Persicaria wallichii* var. *pubescens,* and from E. Cornwall (v.c. 2) in error for 3 *Persicaria campanulata* (Conolly 1991). A native of China occasionally cultivated in British gardens, *Polygonum lichiangense* differs from *Persicaria wallichii* var. *pubescens* in the dense, grey, woolly tomentum of crisped hairs on the lower surface of the leaves and its creamy-white perianth-segments 3–5 mm. Confusion probably arose because Stearn (1969), Akeroyd (1989) and other authors treated *P. lichiangense* as a variant of 3 *Persicaria campanulata,* previously called *Polygonum campanulatum* var. *lichiangense* (W.W. Smith) Steward, a view firmly refuted by Conolly (1991).

Persicaria campanulata 3 (A)
P. mollis (p.42) 4 (B)

4 Persicaria mollis (D. Don) H. Gross

Soft Knotweed
(Illustration on p.41)

Polygonum molle D. Don; *Aconogonon molle* (D. Don) H. Hara

A subshrubby, rhizomatous, softly pubescent perennial, with erect, hollow, rounded stems 100–250 cm, branched above. Leaves 5–22 × (1.5–)2–9 cm, lanceolate or elliptic-lanceolate, tapering at both ends, with sinuate margins and with dense appressed hairs on both surfaces; petiole up to 1 cm, silky-villous. Ochreae up to 5 cm long, but truncate after fall of tip, entire, covered with long silky hairs. Inflorescence a pyramidal panicle 15–30 cm, leafy, with hairy branches 8–10 cm bearing small funnel- or trumpet-shaped, bisexual, homostylous, unscented flowers. Perianth-segments 5, 1.5–3 mm, all of equal width, white, cream or green. Perianth fleshy and purplish-black in fruit, forming an unusual berry-like structure enclosing the trigonous nut, but fruits have very rarely been reported in the wild in Britain. $2n = 20$ (Jaretzky 1928), 32 (Kurosawa 1971). Flowering from June to September.

A native of the Himalayas, grown in British gardens since at least 1840 and first recorded in the wild in 1921; known as an established escape only at Loch Eck, Main Argyll (v.c. 98). Also apparently escaping from gardens or perhaps planted near Beaulieu, S. Hants (v.c. 11), and near Tunbridge Wells, W. Kent (v.c. 16).

This is a variable species. **Persicaria rudis** (Meisn.) H. Gross (*Polygonum rude* Meisn.), from N. India and Bhutan, which has been treated as a variety of *Aconogonon molle* (var. *rude* (Meisn.) H. Hara), is more coarsely pubescent, with stiff, deflexed hairs on the stems and petioles and less silky ochreae. It is sometimes cultivated in gardens.

The taxonomy of these and other taxa in section *Aconogonon* was discussed by Hong (1992).

5. Persicaria weyrichii (F. Schmidt ex Maxim.) Ronse Decr.

Chinese Knotweed

Polygonum weyrichii F. Schmidt ex Maxim.; *Aconogonon weyrichii* (F. Schmidt ex Maxim.) H. Hara; *Pleuropteropyrum weyrichii* (F. Schmidt ex Maxim.) H. Hara

A coarse, rhizomatous, erect perennial 40–100(–180) cm, with hollow, rounded, sparsely branched stems, roughly hairy below and silky-hairy above. Leaves 5–30 × 2–20 cm, broadly to narrowly ovate, acuminate, cuneate to cordate at the base, entire but wavy on the margins, dull green, rugose and shortly hairy above, with dense whitish woolly crisped hairs ± concealing the surface beneath; petiole 1–9 cm, hairy. Ochreae 3–4 cm, brown, silky-hairy, soon disintegrating. Inflorescence large, dense, pyramidal, softly pubescent, consisting of many-flowered axillary or terminal panicles. Flowers functionally unisexual, female ones homostylous, with shrunken anthers. Perianth-segments 5, 1.5–2 mm, greenish-cream, not enlarging greatly in fruit. Nut much exceeding the perianth, 5–7 × 4–5 mm, ellipsoid, trigonous, flattened, broadly 3-winged, pale brown, glossy. $2n = 20$ (Doida 1960). Flowering from June to September.

A native of China, Japan and the Kurile Islands; an occasional escape from gardens in Fennoscandia (Karlsson 2000). It was introduced to Britain as a garden plant probably in the late 19th century. It is known as an established escape only near Wastwater, Cumberland (v.c. 70), where it was first noticed in 1978, although *Persicaria × fennica*, the hybrid between *P. weyrichii* and *P. alpina*, occurs at one site in W. Yorks (v.c. 63) (see under 1 *P. alpina*).

The taxonomy of this and other species in section *Aconogonon* was discussed by Hong (1992).

The dense whitish, crisped, coarse woolly hairs on the undersurface of the leaves are quite different from those of the whitish or buff felt on the leaves of *P. campanulata*.

6 Persicaria vivipara (L.) Ronse Decr. *Alpine Bistort*

Polygonum viviparum L.; *Bistorta vivipara* (L.) Delarbre

A slender, erect, glabrous perennial, 8–40 cm. Rhizome rather stout, swollen and bulb-like at the base of the flowering stem. Leaves 2–7(–10) × 0.5–1.8 cm, linear-lanceolate, obtuse, tapering at the base, with revolute margins, dark green and slightly shining above, puberulent and dull glaucous beneath, the lower petiolate and sometimes broader and rather abruptly narrowed into the petiole; petiole not winged. Ochreae brown, obliquely truncate, laciniate. Inflorescence cylindrical, slender, 3–10 × 0.5–1.0 cm, the lower part with abundant pyriform purplish bulbils *c*. 4 mm long, in axils of bracts; bracts membranous, caudate, 0.7–2.6 mm apart. Flowers with short pedicels, in upper part of inflorescence only. Perianth-segments 5, 3–4 mm, usually whitish, sometimes pink, rarely red. Longer stamens strongly exserted. Nut rarely produced, shorter than the perianth, 2.5–3 mm, pale brown, glossy. $2n = 77$, *c*. 88 (Sokolovskaya & Strelkova 1938), *c*. 100 (Flovik 1940), *c*. 110 (Löve & Löve 1948), *c*. 132 (Skalinska 1950), 66, 88 and 99 (Engell 1978). Variable numbers of B-chromosomes are present. Flowering from June to August.

Native. In mountainous areas, up to 1210 m on Ben Lawers (v.c. 88), on wet rocks and consolidated screes, and in base-rich grassland and damp flushes. Propagules are carried down to lower levels by streams and the plant is often abundant in montane pastures; it descends to sea-level in N. Scotland. Common in the mountain districts of northern England and Scotland, but in Wales restricted to Snowdonia. Rare in Ireland, only in S. Kerry (v.c. H1), Co. Sligo (v.c. H28), Co. Leitrim (v.c. H29), and Co. Donegal (v.cc. H34–35); the Irish distribution was reviewed by Curtis (1993). Protected in the Republic of Ireland by the Flora Protection Order 1987 (IUCN category: Rare).

Circumpolar, extending south into the higher mountains of Eurasia and in the Rocky Mountains.

The flowers may be all male or all female, or the plant may be monoecious. In spite of numerous insect visitors, seed is rarely set. Reproduction is mostly by the bulbils, which frequently start to grow before they fall from the spike. They are distributed by birds such as Ptarmigan and, after they fall, in streams and by wind.

Morphological and cytological studies of the species were made by Engell (1973, 1978). Law, Cook & Manlove (1983) investigated the production of flowers and bulbils in relation to ecological factors. Söyrinki (1989) studied fruit production and seedling survival.

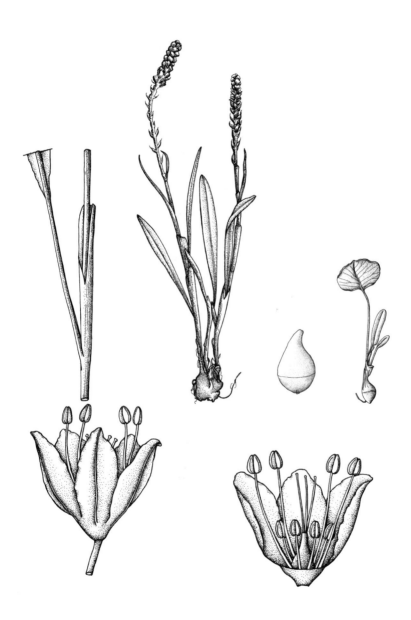

7 Persicaria bistorta (L.) Samp.

Common Bistort, Snake-root, Easter-ledges, Pudding-dock

Polygonum bistorta L.; *Bistorta officinalis* Delarbre; *Bistorta major* Gray

An erect, unbranched perennial 20–120 cm. Rhizome creeping, very stout and fleshy towards the leafy shoots, contorted in knotted lumps, dark chestnut. Leaves mostly basal; basal leaves with lamina 10–25 × 3–10 cm, ovate, obtuse to shortly acuminate, truncate at the base, paler and densely scabrid-puberulent beneath, with petiole up to 30 cm, winged in the upper part; stem leaves sessile, triangular, acuminate, cordate at the base, the uppermost narrow. Ochreae up to 8 cm, obliquely truncate, ± laciniate, brown. Inflorescence cylindrical, dense, stout, 2–6 × 1–1.5 cm, solitary on an erect unbranched stem; flowers scented; peduncle and the short pedicels glabrous; bracts caudate to tricuspidate, *c.* 0.2 mm apart. Perianth-segments 5, 3–5 mm, reddish-pink, rarely white. Longer stamens strongly exserted. Nut a little longer than the perianth, 3.3–4.2 mm, dark brown, glossy. $2n = 44$ (Jaretzky 1928), $2n = 44, 46$ (Sokolovskaya & Strelkova 1938), 24, 48 (Cartier 1938), 48* (Al-Bermani *et al.* 1993, Montgomery *et al.* 1997). Flowering from May to September.

Native. Commoner in upland areas, especially in north-western and north-central England, but scattered throughout Britain (map, p.48); possibly introduced from gardens in much of southern and south-eastern England and in Ireland, where it is local and found mainly in the north. *P. bistorta* forms large patches in damp, humus-rich meadows and pastures, in river valleys and in tall herb communities of alder copses and mountain ledges, usually on neutral or acid soils.

Much of temperate Eurasia; widespread in Europe, but only on mountains in the south, and introduced in many parts of the north. Our plant is subsp. *bistorta*.

This is a popular garden plant, especially cv. 'Superba' with dense spikes of pink flowers. This has received an RHS Award of Garden Merit.

The young leaves of wild plants, known as 'Easter-giants' and 'Easter-ledges', are ingredients in traditional herb puddings served at Easter in Cumbria and the Pennines, especially Calderdale, Yorkshire (Mabey 1996). In the Lake District *P. bistorta* often grows about villages and farm-houses, where it was probably originally planted.

7 Persicaria bistorta

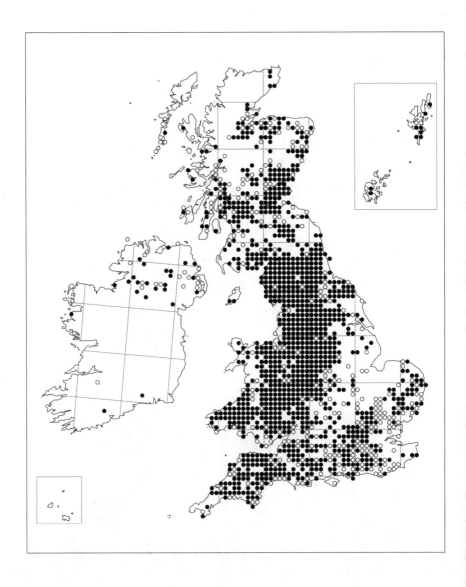

Persicaria amplexicaulis (p.50) 8

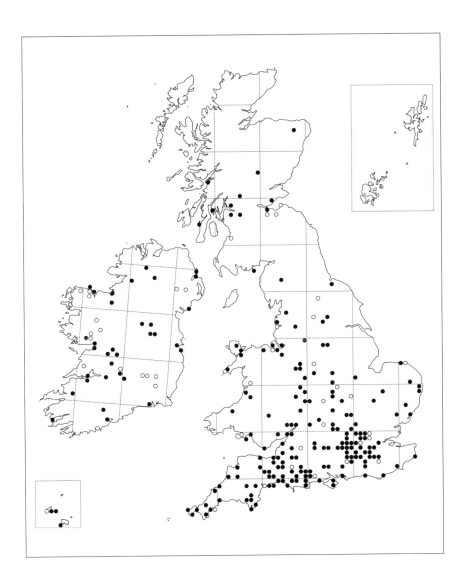

8 Persicaria amplexicaulis (D. Don) Ronse Decr. *Red Bistort*

Polygonum amplexicaule D. Don; *Bistorta amplexicaulis* (D. Don) Greene

An erect, strongly growing, tufted perennial, 50–120 cm, with a branched, woody rootstock, somewhat similar to 7 *P. bistorta*. Leaves 10–25 × 3–9 cm, the lower cordate-ovate to cordate-lanceolate with unwinged petioles, the upper cordate, sessile, amplexicaul, all long-acuminate, scabrid-puberulent beneath. Ochreae large, up to 6 cm, split almost to the base, brown. Spike-like inflorescences one or more per stem, rather loose, 3–8 cm, often in pairs on leafless branches; pedicels *c.* 4 mm; bracts evenly tapering. Perianth-segments 5, 3–6 mm, deep purplish-red or claret-coloured. All stamens slightly exserted. Nut 4–5 mm, pale brown. $2n = 40$ (Mallick 1968). Flowering from July to October.

Native from Afghanistan to S.W. China. Widely cultivated in gardens since its introduction in 1826 and known as an escape since at least 1908, it is now established in scattered localities on roadsides, in hedgerows and on waste ground across Great Britain and Ireland, particularly in southern England and parts of Ireland (map, p.49). It has not been observed to set seed in the wild here and spreads by vegetative means; in gardens especially it can be invasive.

The flower colour distinguishes these naturalized plants as var. *speciosa* (Hook. fil.) Akeroyd (*Polygonum amplexicaule* var. *speciosa* Hook. fil.), whereas var. *amplexicaulis* has red, white or greenish-white perianth-segments. Armitage (2013), who has described taxonomic variation in this species in cultivation, regards var. *speciosa* as insignificant, although the flower colour is consistent for all our naturalized plants.

Persicaria affinis (D. Don) Ronse Decr. (*Polygonum affine* D. Don; *Bistorta affinis* (D. Don) Greene), Himalayan Bistort, a glabrous, mat-forming perennial, with elliptical to linear-oblanceolate leaves, paler beneath, stems up to 30 cm and dense cylindrical inflorescences of pink, reddish or white flowers, is widely cultivated in rock gardens and has occasionally been reported as an escape in the past.

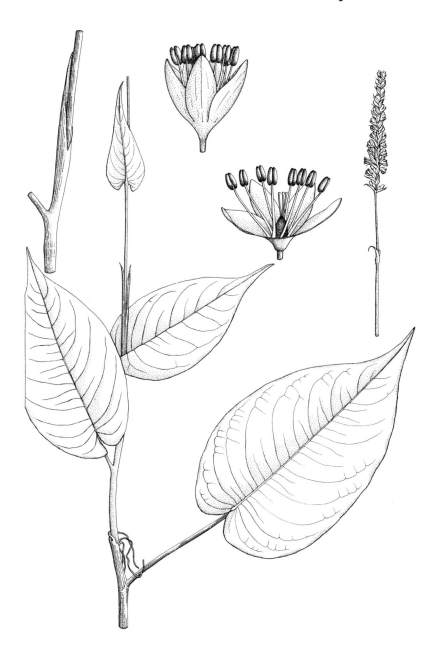

9 Persicaria sagittata (L.) H. Gross ex Nakai   *American Tear-thumb*

Polygonum sagittatum L.; *Truellum sagittatum* (L.) Soják

A weak sprawling annual, 50–100(–150) cm, supported by other vegetation when young; stems 4-angled; stems, petioles, peduncles and leaves with numerous minute, recurved prickles ranging from stiff on the leaf blades, through rigid on the margins to spine-like on the stems. Leaves 3–6 × 1–1.5(–2) cm, ovate or oblong, rather obtuse, sagittate at the base, the lower long-petiolate, the upper petiolate or subsessile; petiole 1–5 cm. Ochreae 5–7 mm, funnel-shaped, very oblique, acute, with stiff hairs. Inflorescences solitary, short, capitate, mainly terminal; peduncle up to 8 cm. Perianth-segments 4, *c*. 4 mm, whitish, pink or greenish. Nut 3–3.5 mm, triquetrous, black or brownish, smooth. $2n = 40$ (Doida 1962). Flowering from July to September.

A native of North America and E. Asia. Its single naturalised European station was by streams and other wet ground about Castle Cove, Kenmare Bay, S. Kerry (v.c. H1), where it was said to have been introduced in 1847 at the height of the Irish potato famine with a cargo of American maize. The evidence for this romantic but unconvincing origin is commented upon by Donaldson *et al.* (1978). The species was first recorded in Kerry by R.W. Scully in 1889, as *Polygonum arifolium* L. (Scully 1890). At the time it grew there abundantly and remained plentiful for many years, but even in the 1970s it was probably on the verge of extinction (Donaldson *et al.* 1978) and was last reported in 1993 (Reynolds 2002). It has also been recorded as a casual at Widdecombe-in-the-Moor, S. Devon (v.c. 3). The Irish plants are referable to var. *americana* (Meisn.) Miyabe.

The closely related **Persicaria arifolia** (L.) K. Haraldson (*Truellum arifolium* (L.) Soják), Halberd-leaved Tear-thumb, from eastern North America, was formerly recorded from Leamington, Warks (v.c. 38), but has not been seen after 1930.

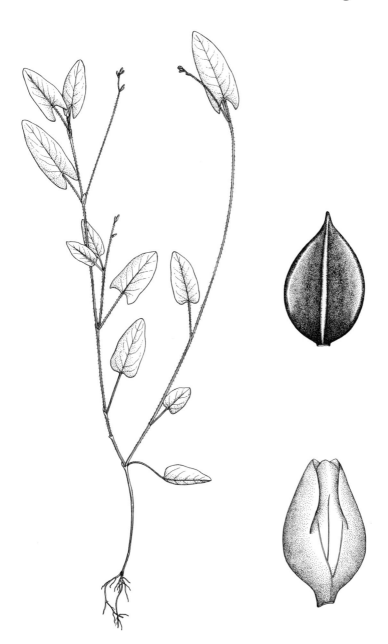

10 Persicaria bungeana (Turcz.) Nakai

Polygonum bungeanum Turcz.; *Truellum bungeanum* (Turcz.) D.H. Kent

An erect, branched annual, 20–60 cm; stems with a few distant, recurved prickles. Leaves 5–12 × 1–5 cm, ovate-lanceolate, acute to subobtuse, cuneate at the base, shortly pilose beneath, especially on the nerves; petiole 5–12 mm, pilose. Ochreae 3–8 mm, membranous, slightly pilose, ciliate. Inflorescence 2–10 cm, branched, lax; peduncle rough with dense, stipitate glands. Flowers on very short pedicels. Perianth-segments 5, 2.5–4 mm, elliptical, pale green or white, slightly red-tinged. Nut 2.5–3 mm, lenticular, brown to black, smooth, dull. $2n$ = 20 (Probatova & Sokolovskaya 1989).

A native of Korea, northern China and Manchuria; reported as an adventive in Japan, Midwestern USA and several European countries, often imported with soya beans. In Britain it is a very rare casual, probably imported with oil-seed. It has been reported only from Olympia sidings, Selby, S.E. Yorks (v.c. 61) by Sledge (1934).

P. bungeana superficially resembles several species in *Persicaria* section *Persicaria*, but the recurved prickles on the stem, although few in number, clearly show its affinity to 9 *P. sagittata* and other species in section *Echinocaulon*.

Persicaria orientalis (L.) P.L. Vilm. (*Polygonum orientale* L.), Princess-feather, is a native of E. and S.E. Asia sometimes naturalised in C. and S.E. Europe and an occasional garden escape in Britain and Scandinavia. Somewhat similar in habit to *P. bungeana* (although within section *Persicaria*), it is a robust annual up to 150 cm, with stiff, bulbous-based hairs mixed with very long-stalked gland-tipped ones, ochreae often terminated by leaf-like appendages, broadly ovate, acuminate leaves, and red to purplish 5-merous flowers in rather dense, sometimes drooping spikes. $2n$ = 22 (Jaretzky 1928).

It was recorded from Middlesex (v.c. 21) in 1953 (Kent 1975) and on a rubbish-tip at Baildon, Mid-W. Yorks (v.c. 64) in 1965 (Clement 1983).

11 Persicaria senegalensis (Meisn.) Soják

Polygonum senegalense Meisn.

A stout rhizomatous perennial (only forming small patches in Britain), with erect, glabrous stems up to 250 cm, rooting at the lower nodes, superficially similar to extremely robust plants of 14 *P. lapathifolia*. Leaves 10–25 × 4–7 cm, oblong-lanceolate to lanceolate, long-acute to acuminate, tapered to a short petiole, glabrous except for strigose hairs on the margins and the veins beneath, and dotted with sessile yellowish glands. Ochreae up to 3.5 mm, truncate, not ciliate, brownish. Inflorescence a branched panicle of elongate spike-like racemes up to 7 cm. Peduncles, pedicels and perianths usually dotted with yellowish glands. Perianth-segments 4, 2–3 mm, rose-pink. Nut *c.* 2.5 mm, lenticular, blackish, glossy.

A native of tropical Africa, extending south to the Cape and north to Egypt, Palestine, Crete and Malta, growing in marshes by lakes and rivers; introduced to Britain as a rare casual, mainly as a wool adventive, and not persisting. It has been recorded from S. Beds (v.c. 30) (J.E. Lousley, **RNG**), Worcs (v.c. 37) (Lousley 1955), mid-W. Yorks (v.c. 64) and, more recently, Surrey (v.c. 17), on arable land or as a ruderal. However, R. Wisskirchen, who re-determined Lousley's specimens as *P. lapathifolia*, does not accept that *P. senegalensis* is present in Britain, considering it to have been confused with robust and broad-leaved alien, probably tropical, variants of *P. lapathifolia*. Further study is required, especially as *P. senegalensis* may well appear again in Britain as an alien, perhaps persisting in response to climate change.

P. senegalensis somewhat resembles 14 *P. lapathifolia* and related species, from which it can readily be distinguished by its robust perennial habit, but belongs within section *Amblygonon*. In its native range it varies considerably in hairiness, with Mediterranean plants mostly lanate-hairy.

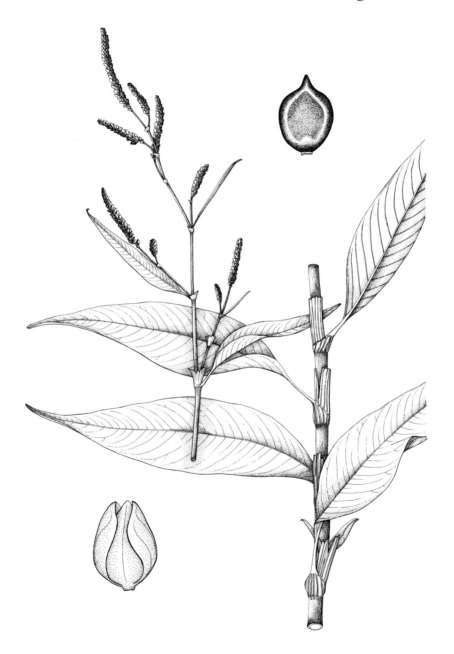

12 Persicaria amphibia (L.) Delarbre

Polygonum amphibium L.

An aquatic or terrestrial perennial with a far-creeping rhizome up to 13 m. Aquatic plants (f. *amphibia*) glabrous and floating, with stems 100–300 cm, producing adventitious roots at most nodes; leaves floating, with petiole 2–4 cm and blade 7–15 × 2–4 cm, ovate-oblong, obtuse, truncate to subcordate at the base. Terrestrial plants (f. *terrestris* (Leers) Kitag.) have erect, little-branched stems 20–100 cm, rooting only at the lower nodes, glabrous or slightly scabrid-pubescent; leaves sessile or very shortly stalked, 7–12 × 1–1.5 cm, lanceolate or oblong-lanceolate, acute or subacute, narrowed to a truncate base, covered with strigose or occasionally glandular hairs and with a ciliate margin. Ochreae acute, hispid when young, ciliate or not. Inflorescence 2–5 cm, obtuse, dense; peduncle stout, glabrous and sometimes glandular. Perianth-segments 5, *c*. 3.5 mm, pink, rarely deep rose-pink, eglandular. Stamens 5, exserted or not. Styles 2. Nut 2–2.7 mm, lenticular, dark brown, glossy. $2n = 66$ (Rudyka 1995), 88* (Partridge 2001), 96* (Al-Bermani *et al*. 1993). Flowering from July to October.

Native and common throughout Britain and Ireland (map, p.60). The aquatic form is found floating in lakes, ponds, canals, slow-flowing rivers and ditches, sometimes covering considerable areas. The terrestrial form grows in damp places on the banks of lakes, canals and rivers and in marshes. It occurs also as a weed of damp cultivated land and grassland, especially on heavy soils, and in periodically flooded dune-slacks. In these drier habitats it often fails to flower. This species is spread mainly by rhizomes, forming extensive clones.

A native of temperate Eurasia south to the Mediterranean region and China, and of North America south to Mexico; naturalized in Central and South America and Southern Africa.

The aquatic and terrestrial phenotypes are so different in appearance that it is sometimes difficult to imagine that they belong to the same species. Otherwise *P. amphibia* had been thought to show little variation in Britain or Ireland, but studies have shown a range of leaf-form and hairiness that probably has some genetic basis. A variant with glandular leaves occurs infrequently in England, and has been reported from Ireland near Bridgetown Priory on the River Blackwater in E Cork (v.c. H5), and formerly at a site in Cork City (v.c. H4).

Grown as an aquatic, this is an attractive and useful ornamental plant.

12 Persicaria amphibia

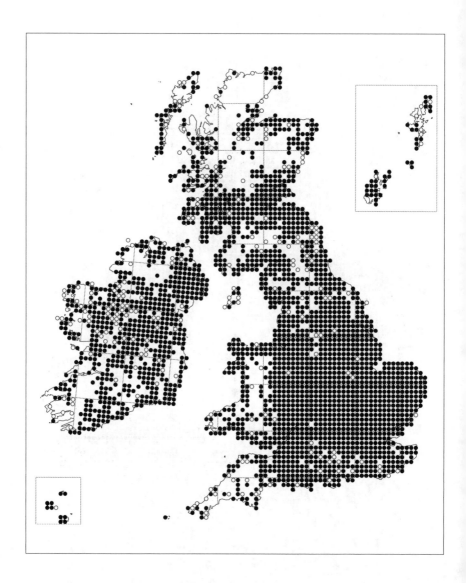

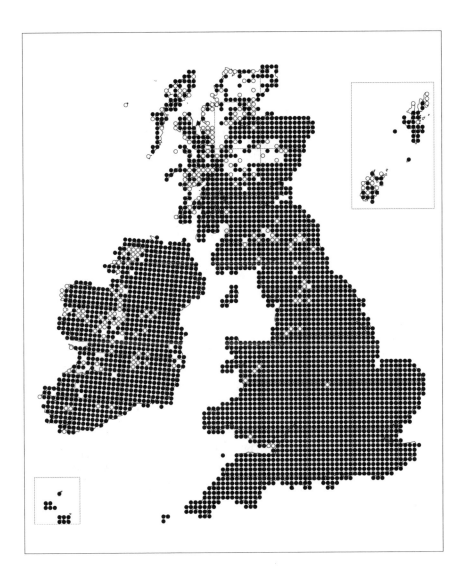

13 Persicaria maculosa Gray   *Redshank, Spotted Persicaria*

Polygonum persicaria L.

A simple or branched, erect or decumbent, almost glabrous annual (occasionally with scattered appressed eglandular hairs), 20–80(–100) cm; stems reddish, swollen above the nodes. Leaves (1–)3–11 cm (up to 15 cm in var. *biformis*), lanceolate or narrowly ovate-lanceolate, acute or acuminate, tapering at the base, sessile or with a petiole not more than 1 cm long, sparsely hairy on the margins and sometimes on the veins beneath, usually with a large black blotch. Ochreae truncate, hairy or glabrous on surface, long-ciliate. Inflorescence rather short, dense, spike-like and usually continuous; peduncle eglandular (rarely with a few subsessile yellow glands), smooth or sparsely appressed-hairy, shining, often pink. Perianth-segments 4(–5), 2.5–3 mm, usually bright pink, sometimes white, eglandular. Stamens usually 5, not exserted. Styles 2 or 3. Nut 2–2.5(–2.8) mm, lenticular and biconvex or bluntly trigonous with concave faces, black, glossy. $2n = 22^*$ (Al-Bermani *et al.* 1993), 44 (Jaretzky 1927). Flowering from June to October.

Native and common throughout Britain and Ireland (map, p.61). By ponds, lakes and streams, in waste places, by roadsides and railways, especially on nutrient-rich soils, and as an all too common annual weed of cultivated land.

Native in temperate Eurasia and N.W. Africa; naturalised in North America.

A variable species; our plants are subsp. *maculosa*. A native variant on drying mud of the banks of rivers and margins of ponds and lakes is simple or little-branched. As a weed on cultivated land, plants are usually much larger, with broader leaves, more branched and with numerous more elongated spikes (var. *biformis* (Wahlenb.) Akeroyd; var. *elata* (Gren. & Godr.) D.H. Kent nom. illegit.). In dry places such as roadsides, railway-tracks, rubbish-tips and waste ground plants are often shorter and more divaricately branched, sometimes prostrate, with leaves 1–4 cm and shorter spikes broader in relation to their length (var. *ruderalis* (Meisn.) Akeroyd).

Persicaria maculosa subsp. **hirticaulis** (Danser) S. Ekman & T. Knutsson (*Polygonum persicaria* subsp. *hirticaule* Danser), from E. Asia, with stiff, spreading, simple, eglandular hairs on the stem and short-stipitate glands on the peduncles, has been reported as a casual at Selby, S.E. Yorks (v.c. 61) (Sledge 1934). There is a specimen from Ireland collected by E.S. Marshall in **CGE**.

Hybrids
Persicaria maculosa occasionally forms hybrids with 14 *P. lapathifolia*, 17 *P. hydropiper*, 18 *P. dubia* and 19 *P. minor*: see under these species.

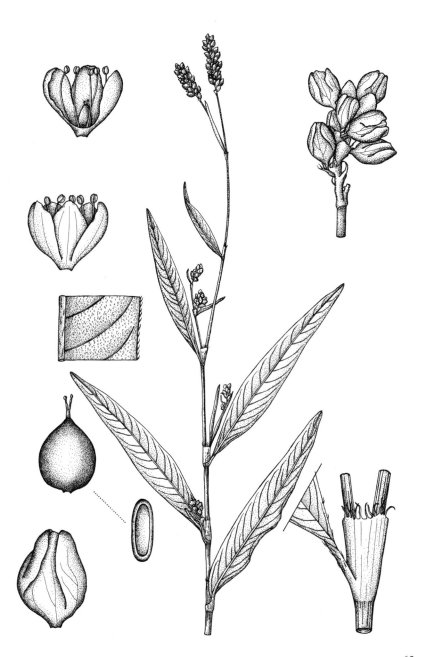

14 Persicaria lapathifolia (L.) Delarbre   *Pale Persicaria*

Polygonum lapathifolium L.

A branched, erect or procumbent, often pubescent annual 30–100 cm, with usually greenish stems, often swollen above the nodes. Leaves 5–15 cm, linear- or ovate-lanceolate, acute or acuminate, tapering to the base, ciliate, sometimes white-lanate beneath, often with a large black blotch; petiole short, 0.5–1.5 cm. Ochreae truncate, the upper ciliate. Inflorescence short and dense, spike-like; peduncle rough with subsessile yellow glands. Perianth-segments 4, *c.* 3 mm, dull pink or greenish-white, densely (subsp. *pallida* (With.) S. Ekman & T. Knutsson) or sparsely yellow-glandular. Stamens 3–5, not exserted. Styles 2. Nut 2–3.5 mm, usually suborbicular, flattened, biconcave, sometimes ± trigonous, black or dark brown, glossy. $2n = 22^*$ (Dempsey *et al.* 1994). Flowering from June to November.

Native. On cultivated and waste ground, river gravels and the drying mud of lakes, ponds, streams and rivers. Common throughout the greater part of Britain and Ireland, though rare over much of northern Scotland and western Ireland (map, p. 67).

A native of temperate Eurasia, N.W. Africa and North America.

A very variable species in which considerable genetic variation is combined with phenotypic plasticity. Several ecotypic variants exist and numerous intraspecific taxa have been described, but patterns of variation in the species (as with other weeds) have been obscured by migration and hybridisation. A distinctive variant with tomentose leaves occurs as an erect weed of cultivated ground on peaty soils, with dense spikes of usually greenish-white flowers, or as a prostrate plant on drying mud around ponds, often with pinkish flowers. This has sometimes been treated at varietal rank in Britain, as *P. lapathifolia* var. *tomentosa* auct., non *Persicaria tomentosa* (Schrank) Bicknell (subsp. *pallida* (With.) S. Ekman & T. Knutsson). Plants with leaves white-lanate beneath have been referred to *Polygonum lapathifolium* var. *salicifolium* Sibth., although allozyme (Consaul, Warwick & McNeill 1991) and biometric (Yang & Wang 1991) investigations suggest that this variant is not taxonomically distinct.

Varieties and subspecies of *P. lapathifolia* have been described in detail by Britton (1933), and more recently Wisskirchen (1991, 1995) has reviewed the extensive variation and evolution in this species.

Many British authorities have accepted **Persicaria nodosa** (Pers.) Opiz (*Polygonum nodosum* Pers.) as a distinct species, although for more than a century field botanists have had difficulty in distinguishing the two taxa. *P.*

64

Persicaria lapathifolia 14 (A)
'P. nodosa' (B)

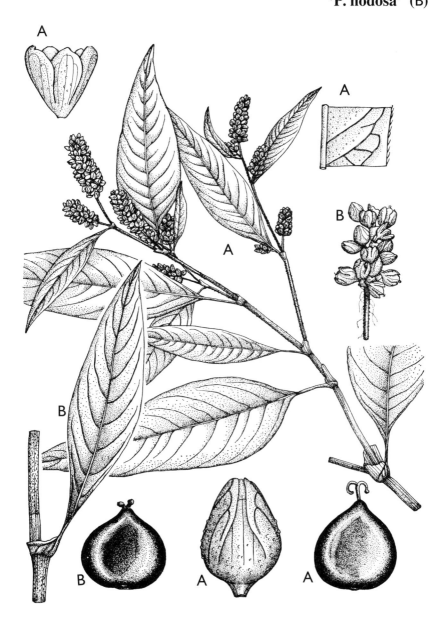

65

nodosa is described as often smaller than *P. lapathifolia*, with a rather lax slender and interrupted spike tapering upwards and ± acute, the peduncles and undersides of the leaves densely dotted with yellow glands, and the nuts only *c.* 2 mm and 'shouldered' at the top. Extreme variants of *P. lapathifolia* and *P. nodosa* are indeed distinct, but a range of intermediates occurs and it seems best to regard *P. nodosa* as part of the considerable variation within *P. lapathifolia* (Timson 1963, Yang & Wang 1991).

Hybrids

Persicaria × **pseudolapathum** (Schur) D.H. Kent (*Polygonum* × *pseudo-lapathum* Schur; *Persicaria* × *lenticularis* (Hy) Soják; *Polygonum* × *lenticulare* Hy) (14 *P. lapathifolia* × 13 *P. maculosa*)

Varies in general form between *P. lapathifolia* and *P. maculosa*. Differs from the former in the reduced glandular clothing of the peduncles and perianths and in the convex and trigonous nuts and from the latter in the glandular peduncles and perianths, the broader fruits, the more prominently veined fruiting perianths and the ochreae with shorter cilia. Apparently very rare, though recorded from S. Devon (v.c. 3), S. Hants (v.c. 11), Surrey (v.c. 17), W. Gloucs (v.c. 34), Mons (v.c. 35) and Glamorgan (v.c. 41). Most herbarium specimens so named are in fact referable to one or other of the putative parents, both of which are very variable (Timson 1975).

Persicaria × *bicolor* (Borbás) Soják (14 *P. lapathifolia* × 18 *P. dubia*) and *Persicaria* × *langeana* (Rouy) Holub (14 *P. lapathifolia* × 19 *P. minor*) have been recorded in Europe.

15 Persicaria glabra (Willd.) M. Gómez   *Hairless Persicaria*

Polygonum glabrum Willd.

A robust, erect, branched annual up to 100 cm. Leaves 5–20 × 2–4 cm, oblong-lanceolate or lanceolate, acuminate with the apex drawn out into a sharp point, glabrous except for a distinctive row of short, yellow, spiny bristles around the margins; lower leaf surface dotted with flat, yellowish glands. Ochreae 2–2.5 cm, membranous with reddish-brown veins, ± entire, not ciliate. Inflorescence a dense, slender-panicled raceme. Peduncles 8–30 cm, glabrous or sparsely glandular. Perianth-segments 5, 2–4 mm, pink or white. Styles 2. Nut 2.5–3 mm, subglobose, lenticular, black, glossy. $2n = 20$ (Mallick 1968).

A native of the Old and New World tropics. Until the 1980s it was a very rare

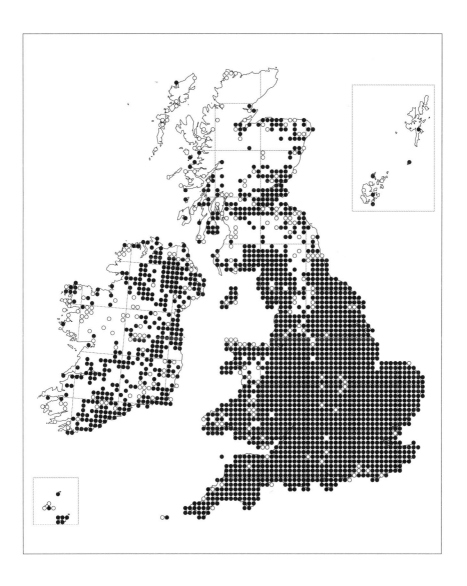

casual in Britain, introduced with wool-shoddy in S. Devon (v.c. 3), but it has turned up here and there in recent years on disturbed ground, probably from bird-seed, always as a casual.

P. glabra can be confused with robust variants of 14 *P. lapathifolia* and the much rarer casual 11 *P. senegalensis*. The leaves of both these species are often dotted with yellow glands; however the yellow bristles on the leaf margins, readily observed with a dissecting microscope, distinguish *P. glabra* from other *Persicaria* species in Britain.

16 Persicaria pensylvanica (L.) M. Gómez

Pinkweed, Pennsylvania Smartweed

Polygonum pensylvanicum L.

A frequently robust, erect annual, branched above, 30–100 cm, with stems usually glabrous below. Leaves 4–20 × 1–5 cm, lanceolate to ovate-lanceolate, long-acute, with sessile orange glands beneath; petiole short. Inflorescence 2–5 × *c.* 1 cm, erect, dense, stout, broad, cylindrical; peduncles usually with abundant divergent stipitate glands, their stalks much longer than their heads. Perianth-segments 5, (2.5–)3–5 mm, rose-pink, eglandular. Stamens 7 or 8. Styles 2 or 3. Nut 3–3.5 × 2.8–3.3 mm, lenticular to trigonous, dark brown to blackish, shining. $2n = 22$ (Löve & Löve 1982). Flowering from August to October.

A native of E. North America that is a sporadic adventive in Europe. First recorded in Britain in 1870; repeatedly introduced with soya beans, oil-seed and bird-seed and reported from waste ground, railway sidings, rubbish-tips and arable fields in widely scattered localities in, for example, N. Somerset (v.c. 6), W. Kent (v.c. 16), Middlesex (v.c. 21), W. Gloucs (v.c. 34), Cards (v.c. 46) and S.E. Yorks (v.c. 61) (Sledge 1934). There is a single Scottish record, from Easterness (v.c. 96); not recorded in Ireland.

A variable species in North America; all our plants seem to have stipitate-glandular peduncles (var. *laevigata* (Fern.) W.C. Ferguson). This handsome plant, with inflorescences superficially resembling those of robust plants of 14 *P. lapathifolia*, appears to be becoming more frequent, although it may be a persistent casual rather than truly naturalised. The stipitate glands, brighter pink flowers and usually slightly larger fruits distinguish it from 13 *P. maculosa*.

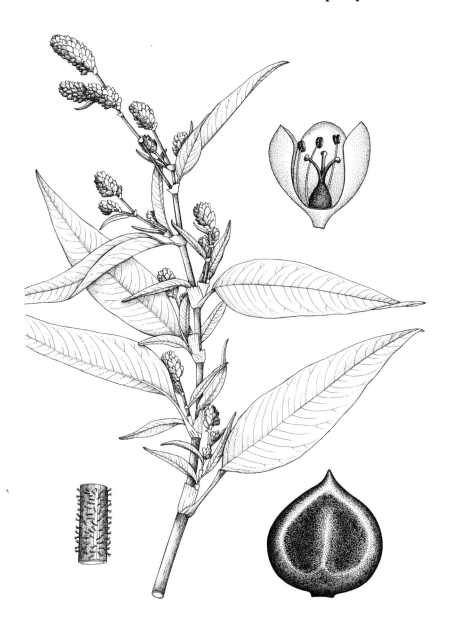

17 Persicaria hydropiper (L.) Delarbre   *Common Water-pepper*

Polygonum hydropiper L.

A nearly glabrous erect or suberect annual 20–80 cm, with often reddish, branched stems, often swollen above the nodes and rooting at the lower nodes. The whole plant, especially the leaves, usually acrid and burning to the taste. Leaves sessile or with a short petiole, 3–9 × 0.8–2 cm, lanceolate, undulate, acute, narrowed to the base, glabrous above, somewhat glandular beneath, asperous on the margins, hispid on the midvein beneath. Ochreae inflated, pubescent or glabrescent, with very short, fine cilia. Inflorescence usually lax, spike-like, slender, nodding, interrupted and leafy in the lower part, the flowers single or in clusters of 2–3 and with a single cleistogamous flower in the lowest leaf-axils; peduncles glandular. Perianth-segments 3–4(–5), 2.5–3 mm, greenish or reddish, covered with flat brownish or yellowish glandular dots, (20–)30–50 on each segment. Stamens usually 4 or 6, not exserted. Styles 2(–3). Nut 2.5–3.5 mm, lenticular or somewhat trigonous, punctate, dark brown or black, dull. $2n$ = 20* (Timson 1966). Flowering from July to October.

Native. In damp muddy places, beside ponds and lakes, by canals, rivers and streams, often in partial shade; usually somewhat calcifuge. Characteristic of shallow depressions where water has stood during the winter but which dry out in the summer, such as vehicle tracks and hoof-marks in woodland rides. Common and widely distributed, although local or rare in parts of northern England and eastern and northern Scotland (map, p.73).

A native of temperate Eurasia, N.W. Africa and North America.

P. hydropiper shows relatively little variation in Britain and Ireland, though sometimes the spikes are stouter, almost cylindrical, and dense-flowered, with flowers in clusters of 3–6 (var. *densiflora* (A. Braun) Akeroyd). This variant is apparently widespread in Britain (Sell & Akeroyd 1989; Akeroyd 2013), especially by rivers and on damp arable land. It occurs in similar habitats in northern Ireland (Hackney 1992).

The peppery-tasting leaves and the cleistogamous flower in the lower leaf-axils of the inflorescence are distinctive features of this species. The glandular perianths and leaves exude a slight sweet scent when bruised (O'Mahony 2003).

The species was monographed by Timson (1966).

Hybrids

Timson (1965, 1966) pointed out that, whilst a number of hybrids involving *P. hydropiper* have been recorded and many herbarium specimens have been

regarded as hybrids, it is doubtful whether hybrids are at all widespread, as *P. hydropiper* is normally autogamous. Determinations of *P. hydropiper* hybrids have been based on morphological characters and have never been confirmed experimentally. In several cases they have included exceptionally large plants that apparently show hybrid vigour. Parnell & Simpson (1989) studied extensive *Persicaria* populations, including hybrids putatively involving *P. hydropiper*, 18 *P. dubia* and 19 *P. minor*, on the shores of Lough Neagh, Ireland, using biometric methods.

Persicaria × intercedens (G. Beck) Soják (*Polygonum × intercedens* G. Beck) (17 *P. hydropiper* × 13 *P. maculosa*)

A branched, rather slender plant, with elongate cilia on the ochreae like those of *P. maculosa*, intermixed with shorter cilia (characteristic of *P. hydropiper*). The elongate, interrupted spikes recall *P. hydropiper*, whilst the large flowers and eglandular peduncles are evidence of *P. maculosa*. Likely to be mistaken for 18 *P. dubia*, but easily separated by the less shining nut. This hybrid has been reported from Surrey (v.c. 17), Berks (v.c. 22), Oxon (v.c. 23), Beds (v.c. 30), Merioneth (v.c. 48) and Derbys (v.c. 57); in Ireland only from Mid Cork (v.c. H4).

Persicaria × figertii (G. Beck) Soják (*Polygonum × figertii* G. Beck; *Polygonum × metschii* G. Beck) (17 *P. hydropiper* × 14 *P. lapathifolia*)

Sometimes intermediate between the parents, but distinguished from *P. hydropiper* by the suberect terminal branches, the shortly ciliate ochreae, the broadly lanceolate leaves, the scarcely glandular peduncles, the ultimately sub-cylindrical, dense-flowered and not or scarcely interrupted spikes, the pink, eglandular perianth-segments and the 5-merous flowers. Very rare and recorded only from Surrey (v.c. 17), Cambs (v.c. 29) and Hunts (v.c. 31).

Persicaria × hybrida (Chaub. ex St-Amans) Soják (*Polygonum × hybridum* Soják; *Polygonum × oleraceum* Schur) (17 *P. hydropiper* × 18 *P. dubia*)

Distinguished from *P. hydropiper* by the shortly hairy ochreae with longer marginal cilia and the usually 5-merous, weakly glandular perianth; and from *P. dubia* by the less hairy and less glandular ochreae with shorter cilia. Very rare and recorded only from Surrey (v.c. 17) and W. Gloucs (v.c. 34).

Variants of 13 *P. maculosa* have sometimes been mistaken for this hybrid (Timson 1975). A plant of *P. hydropiper* from the shore of Lough Neagh, Ireland, closely approaching *P. dubia* in morphology, was not a hybrid (Parnell & Simpson 1989).

Persicaria × subglandulosa (Borbás) Soják (*Polygonum × subglandulosum* Borbás) (17 *P. hydropiper* × 19 *P. minor*)

Likely to be taken for an enormous *P. minor*, from which it may be distinguished by the leaves being abruptly narrowed at the base and by the slightly glandular outer perianth-segments. From *P. hydropiper* it may be distinguished by the slightly hairy ochreae with longer cilia and the mostly 5-merous flowers. Rare, but recorded from W. Sussex (v.c. 13), Berks (v.c. 22), Worcs (v.c. 37) and Mid-W. Yorks (v.c. 64).

Persicaria hydropiper 17

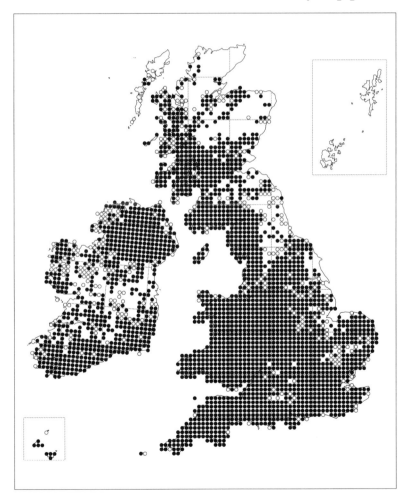

18 **Persicaria dubia** (Stein) Fourr.   *Tasteless Water-pepper*

Polygonum dubium Stein; *Polygonum mite* Schrank; *Persicaria mitis* (Schrank) Opiz ex Assenov; *Polygonum laxiflorum* Weihe; *Persicaria laxiflora* (Weihe) Opiz

A leafy, nearly glabrous, always ± erect annual, somewhat similar to 17 *P. hydropiper* in general appearance but never acrid or burning to the taste, with stems erect from a decumbent base, 30–60 cm, often much-branched, with adventitious roots from the lower nodes. Leaves (30–)50–110 × 12–30 mm, elliptical, broadest at the middle and tapered more or less evenly at both ends, often petiolate, hairy on midvein above and on midvein and lateral veins beneath. Ochreae loose, with appressed hairs, eglandular, with a long conspicuous fringe of cilia. Inflorescence erect, narrow, interrupted, leafy below; peduncle eglandular. Perianth-segments 4–5, *c.* 4 mm, purplish-pink, with 0–15 small colourless glands. Nut 2.8–3.5(–4) mm, blackish-brown to black, fairly shiny. $2n = 40$ (Wulff 1939). Flowering from July to September.

Native. In wet marshy places, shallow ditches and damp hollows on rich soils and beside ponds, lakes and rivers; also in abandoned peat cuttings (map, p. 76). Local and often rare through southern England and Wales, and locally north to Yorks and Cheviot (v.c. 68). In Ireland mostly in Ulster, especially on the shores of Lough Neagh and the adjacent River Bann, where Webb (1984) regarded it as an introduction. However, other recent Ulster records of *P. dubia* from Lough Erne, its discovery in 1998 by the River Shannon in Co. Limerick (v.c. H8), confirming a century-old record (Reynolds 1998), and subsequent records from here and adjacent Co. Clare (v.c. 9) all suggest a wider native Irish distribution.

A native of Europe east from N. Spain, north to northern Ireland, northern England and the Baltic.

P. dubia is frequently confused with 17 *P. hydropiper* but readily distinguished by the more erect inflorescence, the attractive, conspicuous purplish-pink flowers, the lack of cleistogamous flowers and of brownish or yellowish perianth-glands, and the shiny nuts. It can be separated from 19 *P. minor* (especially *P. minor* var. *latifolia*) by its more robust and erect habit, broader leaves and larger nuts. *P. dubia* exhibits little variation in Britain, but in Ireland there is some evidence of introgessive hybridisation along the shores of Lough Neagh, where it occurs in "a rich jungle" of *Persicaria* species (Webb 1984, and see note under *P.* × *wilmsii* below).

Persicaria dubia 18

Hybrids

Persicaria × condensata (F.W. Schultz) Soják (*Polygonum × condensatum* (F.W. Schultz) F.W. Schultz) (18 *P. dubia* × 13 *P. maculosa*)

Stems mostly erect. Leaves lanceolate or oblong-lanceolate, with scattered hairs on the surface, faintly blotched. Ochreae appressed, pubescent, with elongate cilia. Inflorescences numerous, erect, narrowly cylindrical, not interrupted; peduncle eglandular. Perianth-segments rather large, pink, eglandular. Rare and recorded only from N. Hants (v.c. 12), Surrey (v.c. 17), Berks (v.c. 22), Oxon (v.c. 23) and Mons (v.c. 35).

Persicaria × hybrida (Chamb. ex St-Amans) Soják (18 *P. dubia* × 17 *P. hydropiper*): see 17 *P. hydropiper*.

18 Persicaria dubia

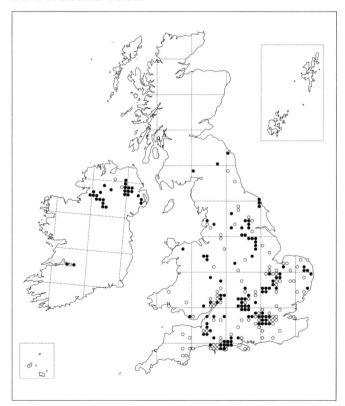

P. × **wilmsii** (G. Beck) Soják (*Polygonum* × *wilmsii* G. Beck) (18 *P. dubia* × 19 *P. minor*)

Superficially resembles a large *P. minor* (but note *P. minor* var. *latifolia*). Leaves scarcely narrowed at the base, resembling those of *P. minor* but broader. Inflorescence narrowly cylindrical, very lax, interrupted, often ± nodding, resembling that of *P. dubia* but with fewer flowers. Perianth-segments pink. Nut *c*. 3 mm, rarely formed. Rare and recorded only from Berks (v.c. 22) and Oxon (v.c. 23); also on the shores of Lough Neagh, Ireland, in Tyrone (v.c. H36) and Co. Antrim (v.c. H39) (Parnell & Simpson 1989).

Using biometric methods, Parnell & Simpson (1989) studied the extensive *Persicaria* populations on the shores of Lough Neagh, including several hybrids between *P. minor* and *P. dubia*. Hybrids showed significant reduction of pollen stainability.

Persicaria minor (p.78) 19

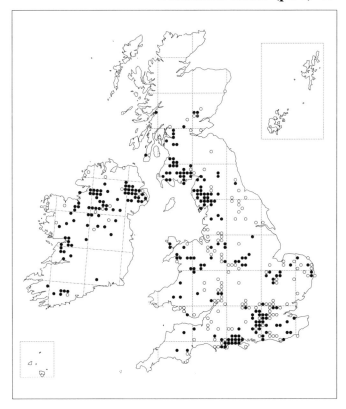

19 Persicaria minor (Huds.) Opiz *Small Water-pepper*

Polygonum minus Huds.

A diffuse, usually much-branched annual with decumbent, rarely erect or unbranched stems 10–30(–60) cm. Leaves sessile, 15–100 × 2–15 mm, linear-oblong to linear-lanceolate, parallel-sided for much of their length, tapered abruptly at the base, rather more gradually at the apex, eglandular, unblotched. Ochreae appressed, eglandular, glabrous or appressed-hairy, conspicuously and coarsely ciliate. Inflorescence lax, slender, spike-like, erect, ± interrupted, narrowly cylindrical. Peduncle glabrous. Perianth-segments 5, *c*. 3 mm, crimson, pink or rarely white, eglandular. Stamens 5. Styles 2(–3). Nut 1.8–2.3 mm, black, glossy. $2n = 40$ (Wulff 1939). Flowering from July to October.

Native and rather rare in wet marshy places, on muddy or gravelly ground beside ponds, lakes and ditches, and in damp places trampled by livestock (map, p. 77). Scattered throughout Britain northwards to Perth (v.cc. 87–89), Moray (v.c. 95) and Main Argyll (v.c. 98); through much of Ireland, except the south-east.

Native in N., W. and C. Europe, and temperate Asia; largely absent from the Mediterranean region; introduced in the Americas.

A variable species. The widespread variant is a small plant with linear-oblong leaves. Plants with sublinear leaves often grow on muddy or gravelly pond margins. At the other extreme are robust plants with broader, lanceolate leaves (var. *latifolia* (A. Braun) Akeroyd) of richer soils in marshy places. These plants have sometimes been confused with 18 *P. dubia* or referred to hybrids. Otherwise *P. minor* is a distinctive, elegant little plant, not usually easily confused with other species. The narrow leaves, slender, elongate, rather interrupted flower-spikes and small nuts should enable positive identification.

Hybrids

Persicaria × brauniana (F.W. Schultz) Soják (*Polygonum × braunianum* F.W. Schultz) (19 *P. minor* × 13 *P. maculosa*)

Usually intermediate between the parents (Roberts 1977). Stem decumbent to ascending, with spreading branches. Leaves blotched, linear-lanceolate or lanceolate, abruptly narrowed at the base. Inflorescence slender, erect, cylindrical, interrupted at the base. Perianth-segments usually large, greenish tinged with purple, red, pink or white. Stamens 6. Usually sterile. Recorded only from Hants (v.cc. 11–12), Sussex (v.cc. 13–14), Surrey (v.c. 17), Berks (v.c. 22), Oxon (v.c. 23) and Anglesey (v.c. 52); in Ireland only from W. and Mid Cork (v.cc. H3–H4).

Persicaria × wilmsii (G. Beck) Soják (19 *P. minor* × 18 *P. dubia*): see 18 *P. dubia*.

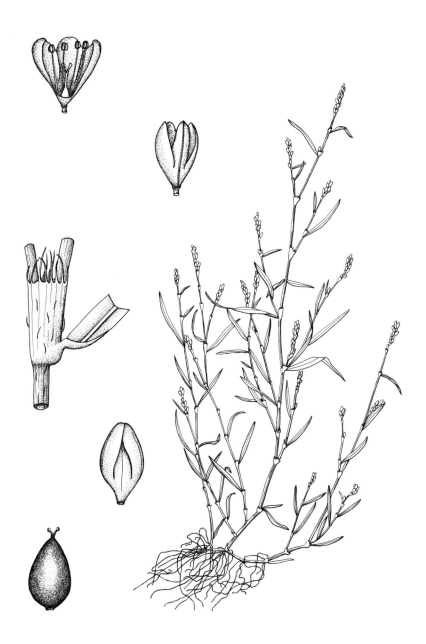

20 Persicaria nepalensis (Meisn.) H. Gross   *Nepal Knotweed*

Polygonum nepalense Meisn.

A straggling annual with a weak stem 10–35 cm, decumbent below and rooting at the nodes, branched, smooth. Leaves 4–9 × 1–3 cm, mostly ovate or elliptical, sometimes with a dark blotch, the lower larger, acuminate, rather panduriform, with the enlarged base auriculate-amplexicaul, the upper sessile, oblong or triangular, acute, cordate at the base; petiole broadly winged. Ochreae caducous, brown, bearing a close tuft of white stiff deflexed hairs at the base. Inflorescences capitate, more or less subtended by an involucral leaf; peduncle with stipitate glands; bracts membranous with a green midrib. Perianth-segments 4(–5), bright pink. Nut enclosed within the persistent perianth, 1.5–2 mm, lenticular, broadly ovoid, minutely punctate, blackish-brown, dull. $2n = 48$ (Doida 1961). Flowering from July to October.

A native of tropical Asia and introduced in Africa, on both continents a weed of cultivation; naturalised in N. Italy and a casual elsewhere in W. Europe, also in North America. A rare or overlooked alien in Britain, first recorded in 1963 and known to be a component of cage-bird seed; it is increasingly widespread and persistent, especially as a weed of nurseries. First recorded on a rubbish-tip at Yeovil, it has occurred, for example, as a nursery weed at Triscombe, S. Somerset (v.c. 5) (Green, Green & Crouch 1997), in nurseries and gardens near Dorchester and at Wimborne, Dorset (v.c. 9), at Hilliers Arboretum, S. Hants (v.c. 11), and at the Royal Botanic Garden, Edinburgh, Midlothian (v.c. 83).

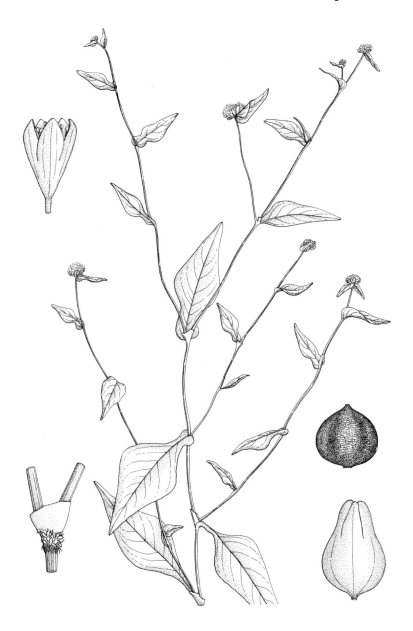

21 Persicaria capitata (Buch.-Ham. ex D. Don) H. Gross

Pink-headed Knotweed

Polygonum capitatum Buch.-Ham. ex D. Don

Somewhat similar 20 *P. nepalensis* but a glandular-pubescent perennial, often forming mats, with weakly ascending, creeping or trailing stems up to 20 cm; stems rooting at the lower nodes. Leaves shortly petiolate, 2–5 × 1–3 cm, ovate to elliptic, acute, often reddish, with a dark, v-shaped stripe. Ochreae densely pubescent, without tuft of white hairs at the base. Inflorescences without a subtending leaf; peduncle 1–3 cm. Perianth-segments 5, pale pink or whitish. Nut 2–2.5 mm, tuberculate. $2n = 20$ (Mallick 1968). Flowering from July to October.

A native of the Himalayas; only half-hardy, it is naturalised in the mild climate of Madeira, the Azores and N. Portugal. It is widely grown in Britain, often in planters or hanging baskets, sporadically escaping and persisting locally for a while, especially at the base of walls and in cracks of pavements in towns and around pubs (Hanson 2002), mainly in milder districts. It also occurs as a weed of nurseries. Until the 1990s it seems to have occurred only in coastal areas and on the Chelsea Embankment, Middlesex (v.c. 21) (Clement 1983), but there are now records from Guernsey, E. Cornwall (v.c. 2) and E. Kent (v.c. 15) north to the Isle of Man (v.c. 71), Cheshire (v.c. 58) and N.W. Yorks (v.c. 65) in a total of 32 vice-counties, perhaps reflecting recent milder winters as well as an increase in sales from garden centres. It has not yet been recorded in Ireland.

The rather similar but much more robust **Persicaria alata** (D. Don) H. Gross, Devon Vine, with prostrate stems 100–200 cm, acuminate leaves up to 7 × 4 cm and flowers in looser heads than in *P. capitata*, is offered by nurseries as a ground-cover plant. Also from the Himalayas, it is hardier and is likely to escape.

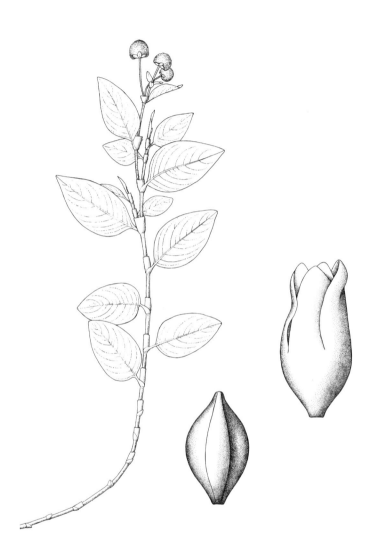

22 Koenigia islandica L.   *Iceland-purslane*

A diminutive, glabrous annual 1–3(–6) cm, with reddish, unbranched or sparsely branched, flexuous stems; the whole plant coloured red in autumn. Leaves 2–5 mm, broadly elliptical to suborbicular, rounded at the apex, rather fleshy. Ochreae c. 1 mm, funnel-shaped, hyaline, brown. Flowers shortly pedicellate, 3–10 in small terminal and upper axillary cymose clusters. Perianth-segments 3, c. 1 mm, broadly elliptical, pale green, often tinged with crimson. Stamens 3(–4). Stigmas 2(–3), capitate. Nut enclosed by the perianth, 1–1.5 × 0.8–1 mm, ovoid, bluntly trigonous, dark brown, ± enclosed by the perianth. 2n = 14 (Mesicek & Soják 1973), 28 (Hagerup 1926; Mesicek & Soják 1973). Flowering from July to September.

Native in open, gravelly flushes and screes at 300–700 m, usually where the ground always remains wet or damp, on the Ardmeanach peninsula on Mull, Mid Ebudes (v.c. 103), and the Trotternish ridge on Skye, N. Ebudes (v.c. 104). Nitrophilous. Although sometimes present in considerable numbers, the plants are sensitive to high temperatures. An apparent decline in recent years is associated with drier springs, and this periglacial arctic plant is extremely vulnerable to further warming of the climate. Fossil evidence indicates a former more widespread postglacial distribution in northern Britain.

This inconspicuous plant was discovered in Britain only in 1934 and then not identified until 1950. Burtt (1950), Ratcliffe (1959), Raven (1952) and Lusby (1999) provide accounts of the species in its Scottish habitats.

Widespread in Arctic regions, including the Faeroes, Iceland and Scandinavia, and mountain ranges south to the Himalayas and W. China; also in the Rocky Mountains and temperate South America (Tierra del Fuego). The Scottish localities are the most southerly in Europe.

It is protected in Great Britain under Schedule 8 of the Wildlife and Countryside Act 1981 (IUCN global category: not threatened; UK category: Near Threatened).

Koenigia is a genus of 6 species, only recently elucidated in full (Hedberg 1997). All the species, with the remarkable exception of *K. islandica*, are restricted to the Himalayas and high mountains of China and S.E. Asia. Ronse Decraene (1989) has discussed the structure of the flowers.

Koenigia islandica 22

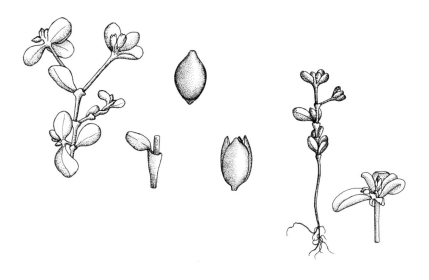

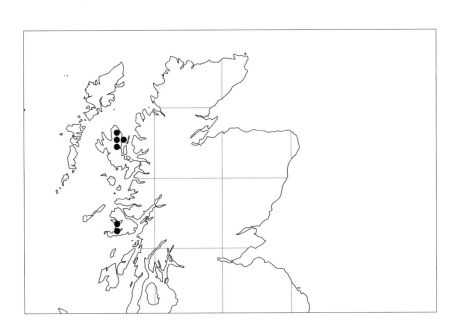

23 Fagopyrum esculentum Moench   *Common Buckwheat*

An often stout, erect, usually little-branched annual 20–100 cm. Leaves 3–8(–10) × 2–5(–7) cm, ovate-triangular, acuminate, cordate, somewhat hairy on the petioles and veins beneath; lower leaves petiolate, the upper amplexicaul. Ochreae 2–6 mm, not fringed. Flowers in short dense axillary panicles on long peduncles, mostly crowded into clusters at the stem apex. Tepals 5, 3–4 mm, elliptical to obovate, cream-white, pale pink or reddish. Nut considerably exceeding the perianth, 5–6 mm, trigonous with slightly concave faces and smooth, entire angles, dark brown, dull. $2n = 16$ (Stevens 1912). Flowering from July to September.

A native of C. and E. Asia, introduced to Britain in bird-seed and pet food and still occasionally cultivated for grain or sown as food for game birds, especially in the south and East Anglia. It is also grown for green manure and for use in herbal medicine. *F. esculentum* occurs sporadically on rubbish-tips and waste ground, in woodland rides, field margins and other places where Pheasants are fed, and as a relic of cultivation.

Buckwheat was probably introduced from Central Asia as a crop during the extensive Tartar incursions into eastern Europe from the 13th century. It has long been an important food crop in Poland, Ukraine and Belarus, growing well on the poorest and sandiest soils. First recorded in Britain in 1597, it was apparently a significant crop in parts of Britain and Ireland during the 17th to 19th centuries.

24 Fagopyrum tataricum (L.) Gaertn.   *Green Buckwheat*

An annual similar to *F. esculentum*, but more slender, with longer ochreae; panicles longer and laxer, arising from the leaf-axils and not crowded into clusters at the apex of the stem; tepals 2–2.5 mm, greenish; nut irregularly rugose, with sinuate-dentate angles. $2n = 16$ (Jaretzky 1927). Flowering from July to September.

A native of Asia; cultivated in various parts of Europe and Asia though much less frequently than *F. esculentum*, with which it sometimes occurs as a weed. An uncommon casual in Britain, first reported in 1961 and mostly introduced with grain, oil-seed and bird-seed, on docksides, railway sidings and rubbish-tips. It generally fails to persist and is probably extinct in many vice-counties.

The species is self-fertile and the flowers may be cleistogamous, in contrast to *F. esculentum*, in which they are frequently cross-pollinated (Chen 1999).

Fagopyrum esculentum 23 (A)
F. tataricum 24 (B)

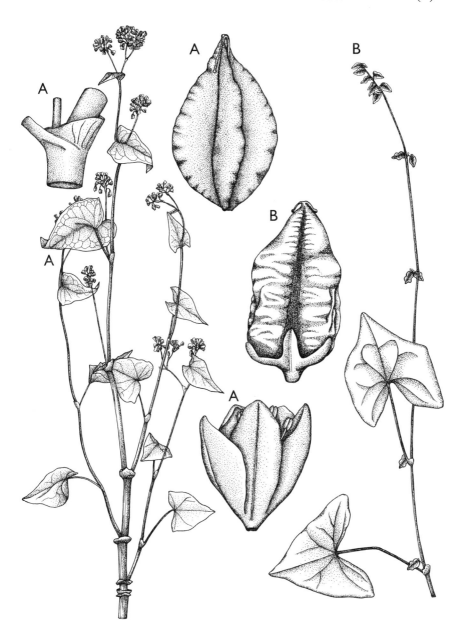

25 Fagopyrum dibotrys (D. Don) H. Hara   *Tall Buckwheat*

A tall, branched, sparsely pubescent perennial with stout stems up to 120 cm. Lower leaves *c.* 10 × 9 cm, broadly triangular, acute or obtuse, cordate, sparsely pubescent; upper leaves narrower, acute, sometimes amplexicaul. Panicle open, with widely spreading branches, each 4–10 cm. Flowers congested near tips of branches, mostly on short pedicels. Tepals 5, 2–3 mm, white. Nut 6–8 mm, ovate, with entire angles. $2n = 16$ (Jaretzky 1928). Flowering from July to September.

A native of the Himalayas and mountains of S.E. Asia. Though little grown in gardens today, it was introduced into Britain before 1846 (Lindley 1846). It is naturalised at a single locality in Wales, just north of Dale, Pembs (v.c. 45), where it is established on a roadside (Lousley 1976). The only other naturalised plants of this species reported from Europe occur near Brest in Brittany, France.

26 Oxygonum sinuatum (Hochst. & Steud. ex Meisn.) Dammer

Awayo

A more or less glabrous annual with decumbent to ascending stems 10–20 cm. Ochreae tubular, fringed with fine setae. Leaves 2–5 × 1–2 cm, ovate to lanceolate, sinuously lobed; petiole 1–2 cm. Flowers 2–3 mm, distant in slender axillary spike-like racemes; bracts small. Tepals 2–3 mm, pink or white, those of male flowers 4–5, fused at the base; perianths of hermaphrodite flowers tubular, 4- to 5-lobed. Stamens 8. Styles 3. Fruiting perianth 5–6 mm, enclosing the fusiform nut, with 3 stout, spreading spines 1–2 mm arising near the middle. $2n = 52$.

A native of E. and C. tropical Africa, Southern Africa, Arabia and Socotra, where it is a weed of disturbed ground. In Britain recorded only once, in 1967, in a timber-yard at Seaforth, S. Lancs (v.c. 59) (Clement 1983).

If it turns up again, it is unlikely to be confused with species of any other genus of Polygonaceae, except possibly *Emex*. Perhaps the most succinct description of the distinctive fruiting structures of *O. sinuatum* is that of Jex-Blake (1948): "perfectly horrid seeds [*sic*], three angled, and at each angle a sharp prickle; these get between dogs' toes and are most painful."

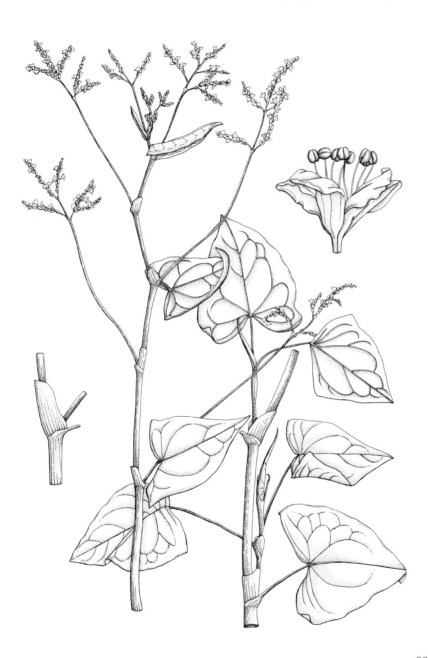

27 Polygonum maritimum L.

A glabrous, prostrate to suberect perennial, woody at the base, glaucous and slightly fleshy, with branched stems up to 50 cm. Leaves 5–25 × 3–8 mm, elliptic-lanceolate, with margins usually revolute, greyish-green, crowded towards the ends of the branches. Ochreae with 8–12 branched veins, silvery-white and conspicuous, usually longer than the upper internodes but much shorter than the lower ones. Tepals 5, *c.* 2–2.5 mm, petaloid, with broad pink or white margins. Nut as long as or slightly exceeding the perianth, 4–4.5 × 2–3 mm, trigonous, chestnut-brown, smooth, somewhat glossy. $2n = 20$* (Styles 1962). Flowering from June to November.

Native. On upper levels of sand, shingle or shell beaches just beyond the reach of the waves except in unusual storms. In such places this perennial species can persist for a number of years but may eventually be destroyed by exceptionally severe climatic or wave conditions. It is also vulnerable to trampling.

A native of the coasts of the Mediterranean, Black Sea and W. Europe north to southern Britain and Ireland. It is notably a plant of Mediterranean seashores, where it can be much larger, almost subshrubby, with the stems forming great 'ropes'. It is also apparently physiologically active there during winter, starting to flower in April. Beeftink (1964) gives an account of the ecology of the species.

The beaches of south-eastern Ireland and the English Channel, with stations in the Channel Islands, Brittany and the Cotentin Peninsula, and one in the Netherlands, are the northern limit of the range of *P. maritimum*. In the past it was recorded from Devon (v.cc. 3 & 4), N. Somerset (v.c. 6) and Dorset (v.c. 9). In 1973 it was discovered in Ireland at a single station in Co. Waterford (v.c. H6) (Ferguson & Ferguson 1974), but it has not apparently been seen since 1974.

It now seems to be better established along the south coast of England (map, p.92). By the beginning of the 1990s it was still known from the Channel Islands, where it has been seen in most years since 1961, and two stations in Cornwall (vcc. 1 & 2), where mild winters have enabled it to persist. Then in 1991 it was confirmed at an old station in S. Hants (v.c. 11), soon followed by a second record in 1995. In 1992 it was reported from E. Sussex (v.c. 14) (Harmes 1994) and in 1995 from Wight (v.c. 10) and W. Sussex (v.c. 13). This apparent rapid spread and persistence may be a result of the mild winters and hot summers that have predominated in Britain since the mid-1970s.

Although expanding its range, *P. maritimum* remains a rare plant in Britain (IUCN global category: not threatened; UK category: Endangered; Irish category: Rare). It is protected in Britain under Schedule 8 of the Wildlife and Countryside

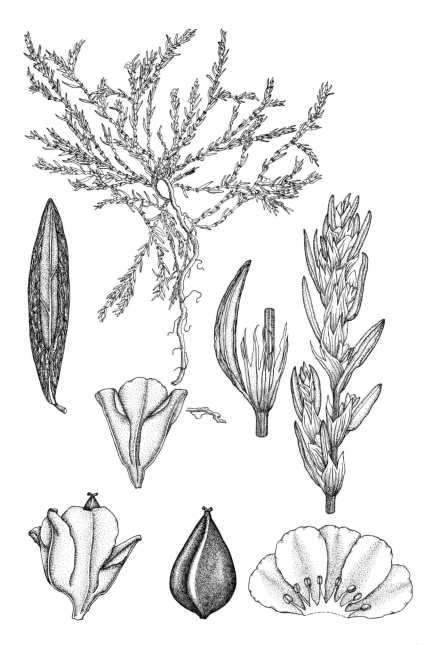

Act 1981 and in the Republic of Ireland by the Flora Protection Order 1987. Threats include the development of beaches and pressure from summer visitors.

P. maritimum can be distinguished from robust specimens of 29 *P. oxyspermum* subsp. *raii* by the distinctly perennial habit, ochreae with 8–12 veins (4–6 in *P. oxyspermum*) and the somewhat smaller nuts. Herbarium specimens of *P. maritimum* frequently blacken on drying.

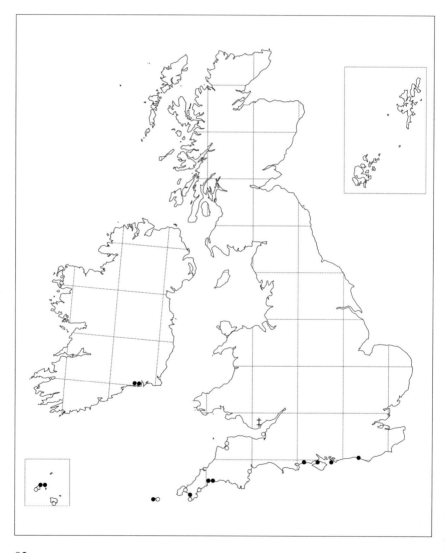

Polygonum oxyspermum subsp. raii (p.94) 28

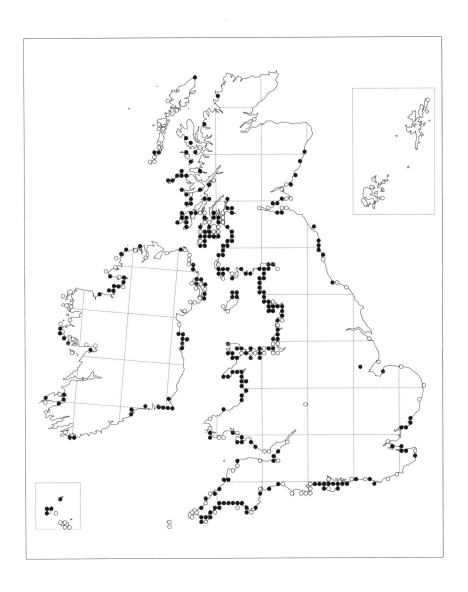

28 Polygonum oxyspermum C.A. Mey & Bunge ex Ledeb.
subsp. raii (Bab.) D.A. Webb & Chater   *Ray's Knotgrass*

A prostrate, glabrous annual, biennial or short-lived perennial, usually glaucous and slightly fleshy, with stems up to 100 cm, becoming somewhat woody at the base. Leaves all more or less equal in size, 10–35 mm, elliptic- to linear-lanceolate, flat, crowded towards the apex of the branches. Ochreae *c*. 5 mm, with 4–6 unbranched veins, shorter than the upper internodes, laciniate, hyaline and silvery in the upper part. Tepals 5, 3–4 mm, conspicuous, green with broad pink or white margins. Nut distinctly exceeding the persistent perianth, 3–5.5 × 3–3.5 mm, trigonous, pale or dark brown, smooth, glossy. $2n = 40*$ (Styles 1962). Flowering from June to September.

Native. Local on shores of sand, shingle or shell debris, sometimes on open sandy ground near the sea, usually growing at the limits of extreme high tides, associated with *Atriplex glabriuscula, Cakile maritima, Euphorbia paralias, Salsola kali* and other plants of strand-lines. Scattered around the coasts of Britain and Ireland, but rare in the east and probably decreasing in some districts owing to increased disturbance of beaches (map, p.93). It is difficult to monitor, however, as, like many other maritime species (Webb & Akeroyd 1991), its presence is erratic and appearances from year to year depend on the influence of storms in turning over beaches. It is unlikely that *O. oxyspermum* subsp. *raii* has appeared every year for any length of time on any one beach. Nordhagen (1963) gives an account of the ecology of the species.

The species occurs on the coasts of N.W. and N. Europe, from Brittany to the Baltic and arctic Norway, on western and central Mediterranean and Black Sea coasts, just extending to the Aegean, and also in N.E. North America.

Subsp. *oxyspermum*, which replaces subsp. *raii* on the shores of the Baltic and southern Norway, has been persistently recorded from Kirkcaldy, Fife (v.c. 85), and sporadically from E. Lothian (v.c. 82). Plants from Arran (v.c. 100) and elsewhere around the Firth of Clyde, with paler nuts 5–6 mm, are close in appearance to subsp. *oxyspermum*. These Scottish plants need further study.

Perennial plants in Britain have sometimes been referred in error to a third subspecies, subsp. *robertii* (Loisel.) Akeroyd & D.A. Webb, endemic to western and central Mediterranean coasts. This plant has smaller, dark brown nuts and is often distinctly woody at the base. Plants from Aegean and Black Sea coasts, often referred to *P. mesembricum* Chrtek, are indistinguishable from subsp. *raii*.

The large glossy nuts, conspicuously exserted from the perianth, and the glaucous leaves distinguish *P. oxyspermum* from 29 *P. aviculare* and related species. Robust specimens of *P. oxyspermum* resemble 28 *P. maritimum* (p.90).

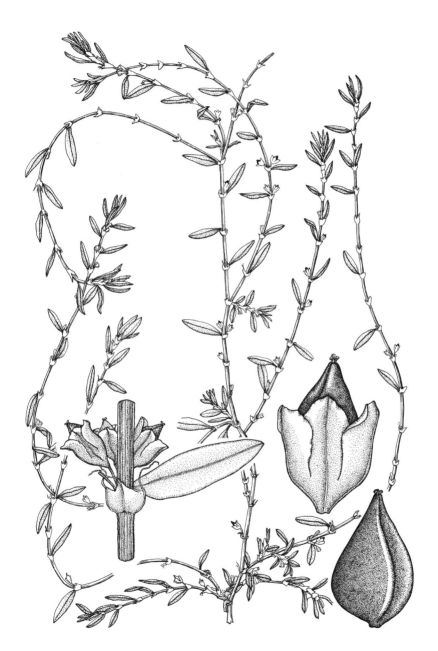

29 Polygonum aviculare L. *Common Knotgrass*

A much-branched, glabrous annual with stems ascending to suberect, sometimes prostrate, up to 200 cm. Leaves of two kinds (plants heterophyllous); those on the main stem larger, 20–50 × 5–18 mm, lanceolate to ovate-lanceolate, sometimes obovate or narrowly elliptical, falling as the plant matures; those on the flowering and other branches about one-third the size, 7.5–15 × 3–5 mm, usually narrower and more persistent; petiole up to 5 mm but usually shorter or absent. Ochreae *c*. 5 mm, brownish at the base, dull silvery-white and laciniate above. Inflorescence of 1- to 6-flowered axillary clusters. Tepals 5, *c*. 2 mm, ovate, subobtuse, greenish, with broad pink or white margins, distinctly veined, united only near the base, overlapping in fruit. Nut included in the perianth, 2.8–4 × 1.5–1.8 mm, trigonous with three equal sides, minutely punctulate, reddish-brown or dark brown, dull. $2n = 60*$ (Styles 1962). Flowering from June to November.

Native. On disturbed and open ground, roadsides and sea-shores, and in arable fields, gardens and waste places. Very common almost throughout Britain and Ireland, except Orkney and Shetland, where it is replaced by 31 *P. boreale* (map, p.98). It can occur in large quantity and remains an injurious weed of agriculture and horticulture.

Native throughout temperate Eurasia and North America, and introduced worldwide as a weed and ruderal.

A very variable species. Several authors, notably Karlsson (2000), have divided *P. aviculare* into a number of subspecies, but we have adopted a conservative view of the complex variation in this species. Further study is needed of British and Irish plants.

Species 29–32 are often recorded as '*Polygonum aviculare* group'. *P. aviculare sensu stricto* may therefore be under-recorded in some areas.

The hybrid between 29 *P. aviculare* and 31 *P. boreale* has been reported in North America and may occur in Scotland.

In autumn, plants of *P. aviculare* are often infested with the powdery mildew *Erisyphe polygoni* and the rust *Uromyces polygoni-avicularis*. Nuts produced from the end of September onwards tend to be elongate and greenish-brown.

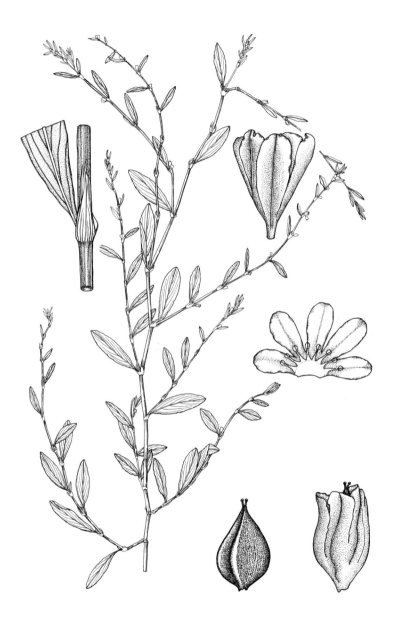

29 Polygonum aviculare

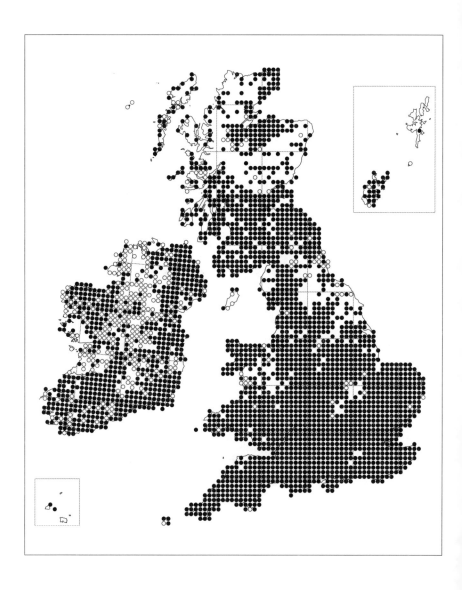

Polygonum rurivagum (p.100) 30

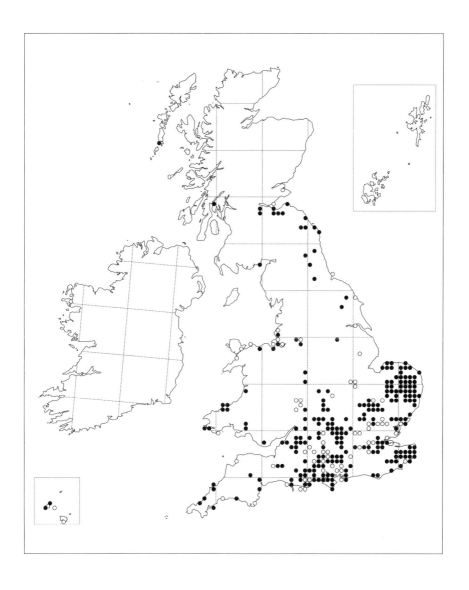

30 Polygonum rurivagum Jord. ex Boreau *Cornfield Knotgrass*

P. aviculare subsp. *rurivagum* (Jord. ex Boreau) Berher

A slender, glabrous annual, usually erect or suberect but sometimes prostrate, seldom more than 30 cm, but stems occasionally up to 60 cm. Leaves of two kinds (plants heterophyllous), the larger leaves on the main stems usually fallen before the fruits appear. Leaves 15–35 × 1–5(–8) mm, linear to linear-lanceolate, rarely lanceolate, acute, subsessile or with a very short petiole; fascicles of smaller leaves often borne on shoots in leaf-axils. Ochreae 5–10 mm, usually longer and more conspicuous than those of 29 *P. aviculare*, reddish-brown below, distinctly silvery above. Tepals 5, *c*. 2 mm, narrow, obtuse, usually red but sometimes pink or whitish, united below but with a distinct gap between the segments above. Nut usually exserted from the perianth, 2.5–3.5 × 1.5–2 mm, trigonous with concave sides, dull. $2n = 60*$ (Styles 1962). Flowering from August to November.

Native. In arable fields, usually cereals, mostly on chalk and other light soils in southern and eastern England, with scattered records north to Kirkudbrights (v.c. 73), Berwicks (v.c. 81) and Fife & Kinross (v.c. 85); more recently found to extend north to Main Argyll (v.c. 98) and Outer Hebrides (v.c. 110), and west to western coasts of Wales (map, p.99). Absent from Ireland, although there are unconfirmed records from Cos Wexford (v.c. H12), Dublin (v.c. H21) and Down (v.c. 38). In Great Britain it is local, but quite common in some areas and undoubtedly overlooked in others.

It is not clear how much this species may have declined or may be under threat from changes in methods of cultivation, although the late flowering season allows seed to be set between harvest and early autumn tillage. *P. rurivagum* sometimes occurs in ruderal habitats, especially at the edge of its distribution, and is apparently being spread with topsoil.

Native in S., W. and C. Europe; precise distribution unknown.

A combination of often erect habit, narrow, acute leaves, reddish, non-overlapping tepals and large nuts distinguishes this species from 29 *P. aviculare* and similar species in the '*Polygonum aviculare* group' (29–32). It is most closely related to *P. aviculare*, with which it shares the same chromosome number, and may represent a segetal ecotypic variant of that species.

31 Polygonum boreale (Lange) Small *Northern Knotgrass*

P. aviculare subsp. *boreale* (Lange) Karlsson

A frequently robust, glabrous annual with stems up to 100 cm, usually erect or suberect, simple or sparingly branched. Leaves of two kinds (plants heterophyllous), 30–50 × 5–20 mm, oblong-ovate to almost spathulate, usually obtuse or subobtuse, with a distinct petiole 4–8 mm. Ochreae 5–8 mm, silvery or brownish. Tepals 5, *c.* 2.5–3 mm, conspicuous, distinctly petaloid, with soft pink or white margins. Nut included within or slightly exserted from the perianth, 3.5–4.5 × *c.* 2.5 mm, trigonous with concave sides, punctate, brown, dull. $2n =$ 40* (Styles 1962). Flowering from late June to October.

Native, in similar places to 29 *P. aviculare*, on roadsides and pathsides, cultivated land, sand-dunes and coastal shingle beaches (map, p.104). Common in Orkney (v.c. 111) and Shetland (v.c. 112), where it largely replaces 29 *P. aviculare* and 32 *P. arenastrum* in the usual habitats of those species; also in E. Sutherland (v.c. 107), W. Sutherland (v.c. 108), Caithness (v.c. 109), Mid Ebudes (v.c. 103), N. Ebudes (v.c. 104) and Outer Hebrides (v.c. 110).

A native of Boreal and Arctic Eurasia, the Canadian Arctic and Alaska.

There are several recent records from Dumfriess (v.c. 72) and Kirkudbrights (v.c. 73) and from around the Firth of Forth, the Isle of May in Fife & Kinross (v.c. 85), Midlothian (v.c. 83) and W. Lothian (v.c. 84); there is also a mid-19th century record from Angus (v.c. 90). These are almost certainly introductions. In Galloway (v.cc. 72 & 73) *P. boreale* has been recorded as a weed of disturbed ground in a vegetable patch and from several road-verges, perhaps as a contaminant of seed mixtures.

The large nuts and oblong-ovate leaves with distinct petioles distinguish this plant from other members of the 29–32 '*P. aviculare* group'.

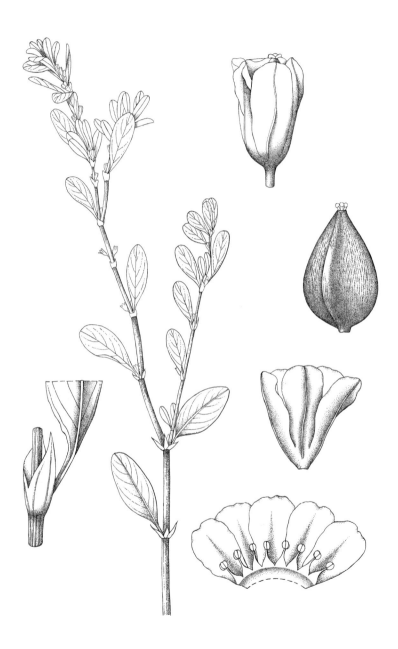

31 Polygonum boreale

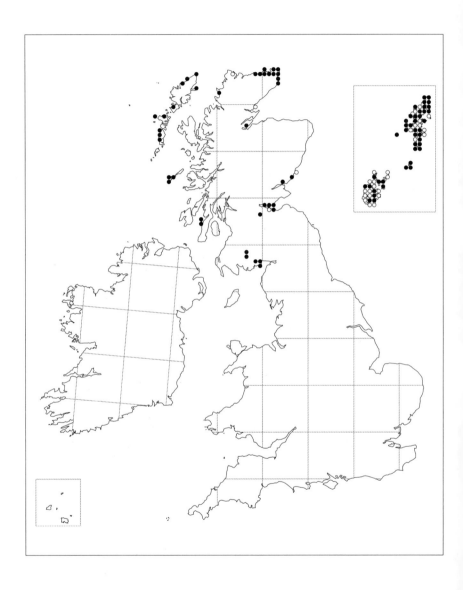

Polygonum arenastrum (p.106) 32

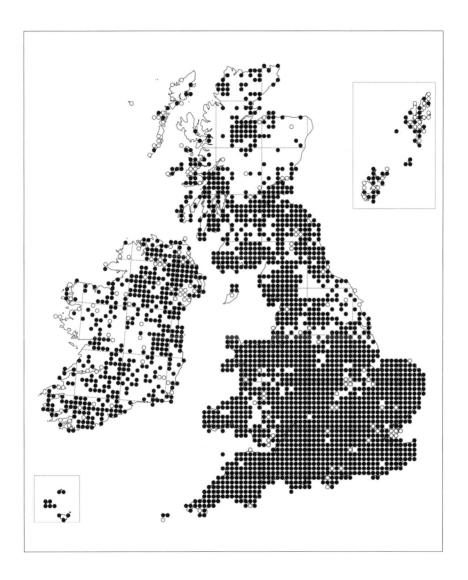

32 Polygonum arenastrum Boreau   *Equal-leaved Knotgrass*

P. aviculare subsp. *microspermum* (Jord. ex Boreau) Berher; *P. aviculare* subsp. *aequale* (Lindm.) Asch. & Graebn.

A much-branched, glabrous annual with stems forming a dense prostrate mat or sometimes weakly ascending, 5–30(–50) cm. Leaves all more or less equal, 8–20 × 2–6 mm, crowded, elliptic to elliptic-lanceolate or lanceolate, subsessile. Ochreae *c*. 4 mm, reddish-brown. Tepals 5, 1.5–3 mm, greenish-white, sometimes pinkish, united for up to half their length. Nut included in or only slightly exserted from the perianth, 1.5–2.5 mm, trigonous, usually with two convex sides and one narrowly concave side, minutely punctulate, medium to dark brown, dull. $2n = 40*$ (Styles 1962). Flowering from July to November.

Native, especially on sand, gravel and other well-drained soils on tracks, paths, drives and playing-fields, in gateways to fields, between paving-stones and on waste ground. It survives in well-trodden places, where it can stand a great deal of trampling under drier conditions than 29 *P. aviculare*. Common through most of Britain and Ireland, except Orkney and Shetland, where it is replaced by 31 *P. boreale* (map, p.105).

A native of temperate Eurasia; also in North America, where it is probably introduced.

P. arenastrum has several ecotypic variants, which show considerable constancy of morphology in relation to habitat. Some authors have regarded these as species. Among the most distinct is a variant with small, narrow leaves and small nuts, *c*. 2 × 1 mm, narrower and more attenuate to the apex than those of typical plants. These plants have been named **P. microspermum** Jord. ex Boreau (*P. calcatum* Lindm.). Styles (1962), Scholz (1977) and others have concluded that they fall within the range of *P. arenastrum* sensu stricto and cannot be treated at specific or subspecific rank. This variant grows on the gravelly margins of ponds and on gravel tracks. In suburban London and elsewhere it is a plant of cracks in pavements and the platforms of railway stations.

In Fennoscandia, Karlsson (2000) distinguishes **P. neglectum** Besser (as *P. aviculare* subsp. *neglectum* (Besser) Arcang.) from *P. arenastrum* on a number of characters, including a more ascending habit, leaves acute or obtuse rather than usually obtuse, pink or red margins to the tepals, and nuts more or less enclosed by the fruiting perianth. Without further research on British and Irish material, it is not yet possible to apply his interesting reworking of the taxonomy of the *P. aviculare* group to these islands. However, a reappraisal of *P. arenastrum* would certainly go much of the way to solving the problem of the small but

Polygonum arenastrum 32 (A)
'P. microspermum' (B)

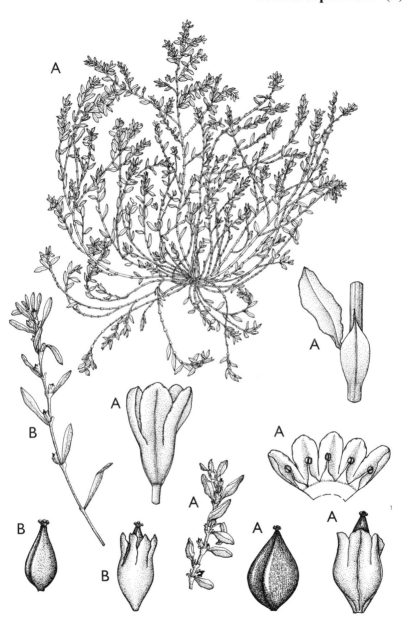

persistent proportion of specimens that is difficult to accommodate within either *P. arenastrum* or *P. aviculare*.

The hybrid *P. arenastrum* × *P. aviculare* (det. C.A.M. Lindman) was recorded from Oxon (v.c. 23) by Druce (1913), but the specimen in **OXF** is *P. aviculare*.

In autumn, plants of *P. arenastrum* are often infested with the powdery mildew *Erisyphe polygoni* and the rust *Uromyces polygoni-avicularis*. Nuts produced from the end of September onwards are often elongate (2.5–4 mm) and greenish-brown.

33 Polygonum cognatum Meisn.   *Indian Knotgrass*

A glabrous perennial with a branched, woody stock and green, prostrate stems, forming patches up to 100 cm or more across. Leaves 5–35 × 2–15 mm, very variable, lanceolate to oblong, subacute or subobtuse. Ochreae hyaline, usually much shorter than the internode but exceeding it in congested variants. Flowers on short pedicels in axillary clusters of 2–5, much shorter than the bracts. Perianth 4–5 mm, pinkish, divided to about halfway into 5 segments, hardening in fruit. Nut included in the perianth, 3 mm, trigonous, black, glossy.

A native of S.W. Asia, locally naturalised in S. France and Bulgaria. Introduced into Britain with grain and formerly found mainly about railway sidings, docks and breweries. It persisted as a ruderal at Westerley Ware, Kew, Surrey (v.c. 17) from before 1872 until 1923; at Colchester, N. Essex (v.c. 19) from 1925 until at least 1961; at Burton-on-Trent, Staffs (v.c. 39) from 1933 until at least 1946, and was then rediscovered there in 2000; and for shorter periods at Bristol, N. Somerset (v.c. 6), Felixstowe Docks, E. Suffolk (v.c. 25) and, at the end of the 19th century, Leith, Midlothian (v.c. 83).

A variable species. Two varieties have occurred in Britain (Lousley 1950, 1953b). *Polygonum cognatum* var. *alpestre* (C.A. Mey.) Meisn. (*P. alpestre* C.A. Mey.) has the pedicel shorter than the calyx and much larger leaves, whereas var. *ammanioides* (Jaub. & Spach) Meisn. has the pedicel the same length as the calyx and leaves much smaller, narrowly lanceolate, acute, with short petioles.

Polygonum cognatum 33 (A)
Polygonum equisetiforme (p.110) 34 (B)
P. plebejum (p.110) 35 (C), P. bellardii (p.111) 36 (D)
P. arenarium subsp. pulchellum (p.111) 37 (E)

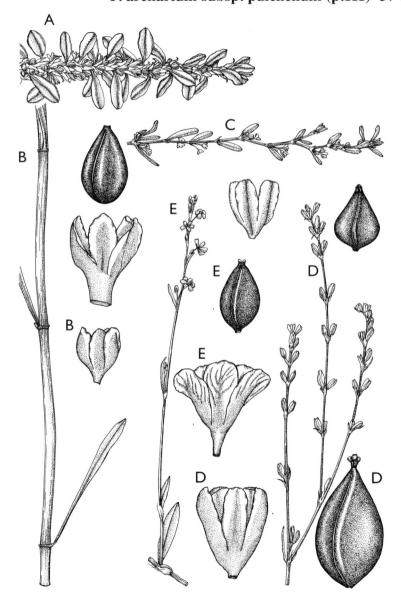

34 Polygonum equisetiforme Sm.   *(Illustrated on page 109)*

A glabrous perennial with a branched woody stock and rather slender, greyish stems 50–200 cm, finely striate, sprawling to more or less erect, with long internodes. Leaves 20–35 × 4–6 mm, oblong to oblong-lanceolate, linear, acute, with prominent veins and rather wavy margins, often caducous. Ochreae 3–4 mm, lacerate. Flowers in clusters of 2–4 in axils of very short scarious bracts, subsessile, forming a loose, rather conspicuous spike at the end of the stem. Tepals 5, *c*. 1.5 mm, pink or whitish. Nut *c*. 2.5 mm, trigonous, minutely punctate, glossy. $2n = 10$ (Weisel 1962), 20 (Castro & Fontes 1946).

A native of open, dry habitats from the Mediterranean region and S.W. Asia east to Afghanistan. A rare casual in Britain, formerly introduced with imported grain or wool and found at docks or in arable fields dressed with wool-shoddy.

The similar **P. scoparium** Loisel. (*P. equisetiforme* auct. hort., non Sm.), endemic to sandy and rocky ground in Corsica and Sardinia, with erect, switch-like stems and entire ochreae, is sometimes grown in gardens (Akeroyd & Rutherford 1987).

35 Polygonum plebejum R. Br.   *Small Knotgrass*
(Illustrated on page 109)

A nearly glabrous prostrate annual resembling the '*P. aviculare* group' (29–32), but smaller and often compact, with much-branched, minutely puberulent stems 6–15(–30) cm, with short internodes. Leaves 6–25 × 1–2.5 mm, narrow, linear-oblong to linear-obovate, obtuse, without lateral veins. Ochreae silvery, with concolorous veins, fringed with laciniae of varying lengths up to 2 mm. Flowers solitary or 2–4(–5) in axillary clusters, pedicellate. Tepals 4–5, 1.5–2 mm, bright pink or greenish. Nut enclosed within the persistent perianth, 1–2 mm, ovoid, broadest near the middle, dark brown or black, smooth, glossy. $2n = 20$ (Sharma & Chatterji 1960).

Native of East Tropical Africa, Egypt, Madagascar, S. Asia and Australia. A variable species that has been introduced occasionally into Britain as a casual wool alien; possibly overlooked or misidentified. The single Irish record, from Limerick docks (v.c. H8) in 1900, was reported as 36 *P. bellardii* (Reynolds 2002).

Polygonum corrigioloides Jaub. & Spach, from C. and S.W. Asia, a glaucous annual somewhat similar to *P. plebejum* (superficially to *Corrigiola litoralis*) was reported at Bristol as a casual before 1930. It has linear-spathulate, obtuse leaves 5–15 mm, flowers in clusters of 2–6, tepals 1–1.5 mm and nuts *c*. 1 mm.

36 Polygonum bellardii All. *Red Knotgrass*
(Illustrated on page 109)

P. patulum auct., non Bieb.

A glabrous, erect annual with a more or less branched stem 20–70 cm. Lower leaves 25–50 mm, lanceolate or oblong-lanceolate, acute, caducous; uppermost leaves much smaller. Ochreae hyaline, laciniate, 6- to 8-nerved, silvery. Inflorescences very lax. Flowers subsessile or on pedicels to 4 mm, solitary or in axillary groups of 2–5; upper bracts very small, scarious. Tepals 5, 2–3 mm, erect, green, with whitish, red, pinkish or purplish margins. Nut 2.5–4 mm, trigonous, brown, glossy. $2n = 20$ (Jaretzky 1928).

Native as a weed of cultivation in S. and W. Europe, extending northwards to N. France, and in North Africa; introduced to Australia. It has occasionally been introduced into Britain as a casual with grain, agricultural and cage-bird seed and wool; found mostly on rubbish-tips, about docks, near mills and in fields where wool-shoddy has been applied. Most records are from the Midlands, northern England and Scotland; it has not been reported in Britain since the mid-1960s. There is a single Irish record, from Tillysburn near Belfast (v.c. H38) in 1939 (Brenan & Simpson 1949).

Raffaelli (1979) has shown that the name *P. patulum* Bieb., long used by European botanists for this plant, correctly refers to another species from the steppes of S.E. Europe and C. Asia, with a more slender habit, flowers crowded towards the end of the branches, and smaller fruits. This other species has never been correctly reported from Great Britain or Ireland.

37 Polygonum arenarium Waldst. & Kit. *Lesser Red Knotgrass*
subsp. pulchellum (Lois.) Thell. *(Illustrated on page 109)*

A spreading, diffusely branched, glabrous annual with flexuous stems 20–70 cm, much-branched from the base, with the branches usually ascending. Leaves 15–30 × 2–6 mm, lanceolate or linear-lanceolate, caducous. Ochreae brown at the base, hyaline towards the apex, 4- to 6-nerved. Flowers in lax spikes, pedicellate, in groups of 1–3, in the axils of the leaves and very small acute, scarious bracts near the apices of the branches. Tepals 5, *c.* 2 mm, conspicuous, patent, pink or white. Nut *c.* 2 mm, minutely tuberculate, dark brown, dull. $2n = 20$ (Löve & Löve 1956).

A native of the Mediterranean region, extending to W. Asia. A rare introduction in Britain since at least 1905; a casual on rubbish-tips, about docks and formerly in fields where wool-shoddy had been applied.

111

An extremely variable plant, which British and other European botanists have sometimes confused with 36 *P. bellardii* and (see note under *P. bellardii*) *P. patulum* Bieb. *P. arenarium* subsp. *arenarium*, from E. Europe and W. Asia, with flowers in more crowded spikes and smooth, glossy nuts, has not been recorded in Britain after 1930.

38 Fallopia convolvulus (L.) Á. Löve   *Black Bindweed*

Bilderdykia convolvulus (L.) Dumort

A somewhat papillose-mealy, prostrate, scrambling or climbing annual with flexuous, twining stems 20–120 cm long. Leaves 2–8 × 1–5 cm, ovate, acute to acuminate, cordate-sagittate at the base, papillose on the petiole and veins beneath; petiole up to 3.5 cm. Ochreae obliquely truncate, more or less laciniate. Inflorescence pedunculate or subsessile, interrupted; pedicels 1–3 mm, jointed above the middle. Perianth-segments 5, greenish-white or pinkish, the 3 outer obtusely keeled or narrowly winged in fruit. Nut 4–5 mm long, dull black, minutely punctate. 2n = 40* (Bailey & Stace 1992). Flowering from July to October.

Native and common throughout Britain and Ireland, although apparently decreasing in northern Britain (map, p.114). A frequent weed of arable land, allotments and gardens, also waste places and roadsides and sometimes on coastal shingle. It is known from the archaeological record to have been a weed of cultivation since the Neolithic. The nuts were formerly the major weed contaminant of agricultural seed in Britain, and they still comprise a significant proportion of the seed-bank in the soil of cultivated land.

A native of Europe and temperate Asia; naturalised in North America and other temperate regions. The species varies considerably in habit and colour, but the variant that attracts the most notice is a plant with broad wings on the fruits (var. *subalata* (Lej. & Court.) D.H. Kent). This is occasional on nutrient-rich cultivated soils in Britain and, more rarely, in Ireland, but is especially a plant of rubbish-tips, where it may be introduced with bird-seed; formerly also on land where wool-shoddy had been applied. This variant is sometimes mistaken for 39 *F. dumetorum*, from which it can be readily distinguished by the larger, dull nuts.

The hybrid *Fallopia convolvulus* × *F. dumetorum* (L.) Holub (*Polygonum* × *convolvuloides* Brügg.), known from C. Europe, has been recorded from Britain, but the records require confirmation.

38 Fallopia convolvulus

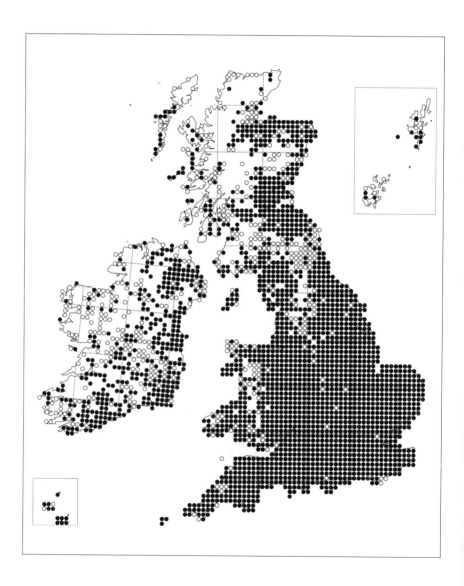

Fallopia dumetorum (p.116) 39

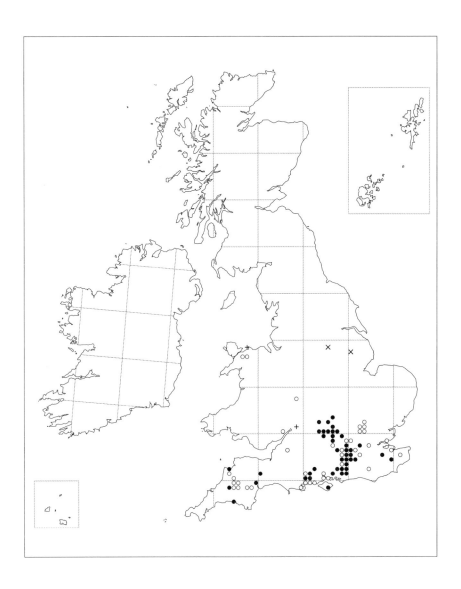

39 Fallopia dumetorum (L.) Holub   *Copse Bindweed*

A climbing annual very similar to 38 *F. convolvulus*, with which it is sometimes confused (especially *F. convolvulus* var. *subalata*), but more robust and with stems up to 3 m or more. Leaves with lamina 3–6 × 1–3 cm, usually thinner and with a more conspicuously acuminate apex than those of F. convolvulus. Inflorescence of interrupted axillary spikes up to 10 cm long; pedicels in fruit up to 8 mm, jointed at or below the middle and deflexed. Outer perianth-segments in fruit much enlarged, c. 8 × 5 mm, broadly and conspicuously winged and decurrent on the pedicel. Nut 2.5–4 mm long, glossy, black. 2n = 20* (Bailey & Stace 1992). Flowering from July to October.

Native and rare in hedges and wood borders on well-drained soils. Erratic in appearance, it is sometimes found in quantity after the felling, thinning or coppicing of hedgerows and woodland. It occurs in central southern England from Dorset (v.c. 9) to Kent (v.cc. 15–16) and formerly extended at scattered sites west to Devon (v.cc. 3–4), north to Worcs (v.c. 37), Warwicks (v.c. 38) and Caerns (v.c. 49). It has always been local but has declined markedly in Britain over the last 30 years. Absent from Ireland.

The species is native across much of Eurasia and is introduced in North America; in Europe it occurs widely north to N. Sweden, although it is rare in the Mediterranean region. *F. scandens* (L.) Holub, a similar species native to North America, has been thought to be be conspecific with *F. dumetorum*, but data from morphology (Kim, S.-T. *et al.* 2000) and flavonoids (Kim, M.-H. *et al.* 2000) suggest that the two species are distinct.

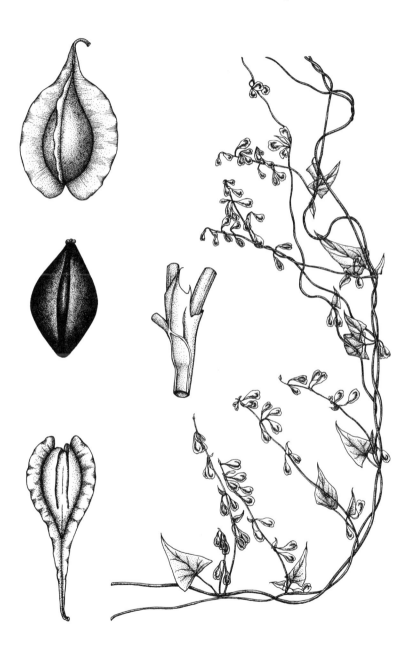

40 Fallopia baldschuanica (Regel) Holub   *Russian Vine*

A vigorous climbing perennial, with tough, woody climbing stems to 12 m or more, forming a mass of twining, vine-like growth over trees and hedges. Leaves bronze-red when young, on long petioles; lamina 3–8 cm, oblong-ovate, ± acute or acuminate, cordate or subcordate at the base. Flowers small, c. 5 mm across, in large diffuse, drooping, branched axillary and terminal panicles, with scabridulous axes, and very showy in the autumn. Perianth-segments white or white tinged with pink, turning pink as the panicle goes into fruit, the outer fruiting perianth-segments with broad decurrent wings. Nut 4–5 mm long, blackish or brown, dull, with finely granular surface. 2n = 20* (Bailey & Stace 1992). Flowering from August to November.

Native to Tadzhikistan (described from Khanate of Baldzhuan), Afghanistan and W. China; naturalised in Spain and C. Europe. In Britain, where it was introduced c.1894, it grows all too well in cultivation, so that roots are often thrown out on to commons, railway embankments, hedge banks, ditches and rubbish-tips. It sometimes conspicuously festoons hedges and scrub. First recorded in 1936, it is now a widespread established outcast in many parts of lowland England and in Wales, but is apparently still rare to local in most of Scotland and Ireland. Spontaneous germination of seed has been observed in Leicester (v.c. 55).

Fallopia aubertii (Louis Henry) Holub, from W. China, is very closely related to *F. baldschuanica* and the two are frequently confused. It has somewhat undulate, shiny leaves and white or slightly greenish flowers that show little pink coloration even in fruit, in smaller, more erect panicles than those of *F. baldschuanica*. It is also cultivated in gardens, although much less frequently, and is only doubtfully naturalised. This plant is probably not specifically distinct from *F. baldschuanica*, the two being perhaps geographical variants of one species.

Hybrids

F. baldschuanica hybridises (as the pollen parent) with 41 *F. japonica* vars *japonica* and *compacta*, 42 *F. sachalinensis* and 43 *F. × bohemica*, although only the cross with *F. japonica* var. *japonica* (*F. × conollyana*) has become established.

Fallopia × conollyana J.P. Bailey (40 *F. baldschuanica* × 41 *F. japonica*)

Similar to 41 *F. japonica* var. *japonica*, having the same rhizomatous habit and fimbriate stigmas, but it has more slender hollow stems less than 1 cm in diameter, bowing over, woody at the base but dying back in winter, narrower

118

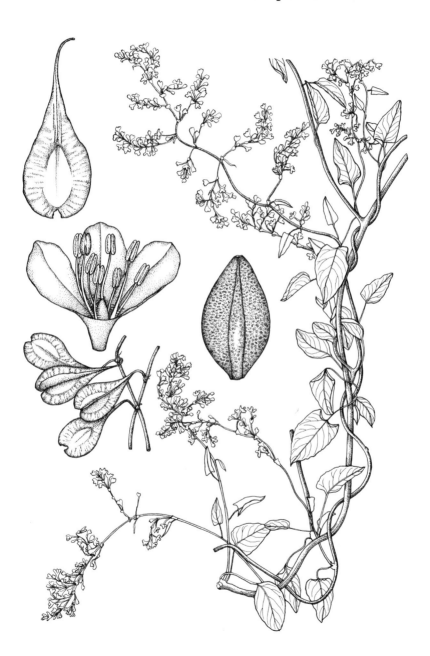

triangular ovate leaves, acuminate to acuminate-cuspidate at the apex, sub-cordate to truncate at the base, and larger (up to 3 mm in diameter), more conspicuous flowers with the 3 keeled perianth-segments much more conspicuously winged. 2n = 54* (Bailey 2001).

This hybrid rarely establishes itself in the wild, despite genetic evidence for a high frequency of seed of this parentage formed by plants of *F. japonica* in Britain. However, a single large, well-established plant has been known since 1987 in an overgrown former railway yard at Haringey, London, Middlesex (v.c. 21) (Bailey 1988, 2001), and more recently some younger plants and seedlings have been found around Haringey and also in a garden in Northants (v.c. 32) (Bailey & Spencer 2003). It has also been reported from the Czech Republic, Germany, Hungary and Norway.

41 Fallopia japonica (Houtt.) Ronse Decr.   Japanese Knotweed

Reynoutria japonica Houtt.

a. var. japonica

Invasive perennial with stout rhizomes, forming extensive, dense clumps. Stems annual, 200–300 cm tall, stout, erect, glaucous, often with reddish spots, branched and flexuous above. Leaves 5–12 × 5–8 cm (length:width ratio 1–1.5), broadly ovate, cuspidate, truncate at the base, coarse, glabrous above and beneath except for minute conical papillose trichomes along the margins; petioles 1–3 cm long. Flowers male-sterile (in Britain and Ireland), 2–3 mm across, creamy-white, in clusters of 3–6 in axillary panicles. Panicles erect at first, drooping at maturity, with the main axis up to 10 cm long with slender branches 6–9 cm long, shorter than the leaves. Perianth-segments 5, the outer 3 keeled. Stamens 8, empty and included within the perianth. Styles 3, exceeding the perianth. Nut 3–4 × 1–1.3 mm, trigonous, dark brown, glossy, enclosed by the enlarged, winged fruiting perianth. 2n = 44 (Sugiara 1931), 88* (Bailey & Stace 1992). Flowering from August to October.

Native of China, Japan and Korea, widely introduced in Europe and North America, also in New Zealand. Now common in the lowlands of Britain and Ireland, up to c. 300 m in the hills, and still increasing (map, p.123). It is especially abundant around towns, in waste places, on roadsides, railway embankments and cuttings, refuse-dumps and mine spoil-tips, along river banks and in places where garden rubbish is dumped. This aggressive garden plant, introduced from Japan probably in the 1850s (Bailey & Conolly 2000), was extensively

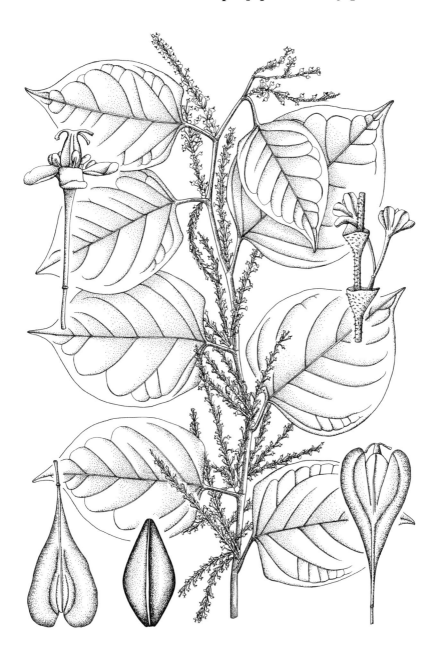

grown by the Victorians. It has been widely established as an outcast in Britain since about 1930 (first reported in the wild in 1886) and is now widespread except in parts of the Highlands and Islands of Scotland. There is little documentation of naturalisation in Ireland before 1939 (Conolly 1977), but it now occurs almost throughout the island.

F. japonica will form clumps wherever rhizomes are dumped, and sometimes, as along the River Lyn in N. Devon (v.c. 4), these are spread by river floods. Several botanists and conservationists have argued that along some urban rivers, for example the River Kelvin in Glasgow (Hart *et al.* 1997) and the River Don in Sheffield (Gilbert 1992), stands of *F. japonica* create a species-rich habitat that mimics deciduous woodland.

The history of the spread of the species in Great Britain and Ireland is given by Conolly (1977) and the history of its introduction is discussed by Bailey & Conolly (2000). It has been monographed by Beerling *et al.* (1994). Molecular data indicate, astonishingly, that *Fallopia japonica* var. *japonica* is represented by a single clone in Britain (Hollingsworth & Bailey 2000). In its native Asia this variety is much more variable; comparisons between the behaviour of the species in Japan and in Europe are given by Sukopp & Sukopp (1988) and Bailey (2003).

b. var. **compacta** (Hook. f.) J.P. Bailey

Reynoutria japonica var. *compacta* (Hook. f.) J.P. Bailey

Plant 50–70(–150) cm tall, with stiffer stems and smaller leaves than var. *japonica*. Leaves 6–8 × 6–8 cm, suborbicular, thicker and more leathery and darker green than those of var. *japonica*, with an abruptly cuspidate apex and undulate margins. Flowers male-sterile or hermaphrodite (gynodioecious), white or more usually reddish, in more erect, unbranched or subsimple inflorescences; male-sterile plants often with conspicuously red stems and fruiting perianths. 2n = 44* (Bailey & Stace 1992).

Native of Japan; a montane ecotype of *P. japonica*, and a more suitable garden subject, sometimes found as an escape. Reported from a number of vice-counties, including E. Cornwall (v.c. 2), but there are few places where the plant is truly naturalised. This variety is discussed by Conolly (1977). In W. Scotland a similar but much larger plant with the leaf lamina c. 10 × 11 cm, almost square in outline, with straightish sides and a small point at the apex, is found occasionally on roadsides. The most spectacular site is in the region of Benderloch and South Ledaig (v.c. 98), where this taller clone has spread extensively along the roadside.

Hybrids

Because of male sterility, all seed formed by *Fallopia japonica* var. *japonica* in Britain is of hybrid origin. Hybrids occur between var. *japonica* and 42 *F. sacha-linensis*, giving 43 *F.* × *bohemica* (2n = 44), between var. *japonica* and 40 *F. baldschuanica*, giving *F.* × *conollyana* (2n = 54: see under 40), and between vars *japonica* and *compacta*, giving the so far unique 2n = 66 hybrid found at Buryas Bridge, Cornwall (v.c. 1) (Bailey & Conolly 1991). Pollination of var. *compacta* by *F. baldschuanica* also produces viable progeny but no plants have yet been recorded from the wild.

Fallopia japonica 41

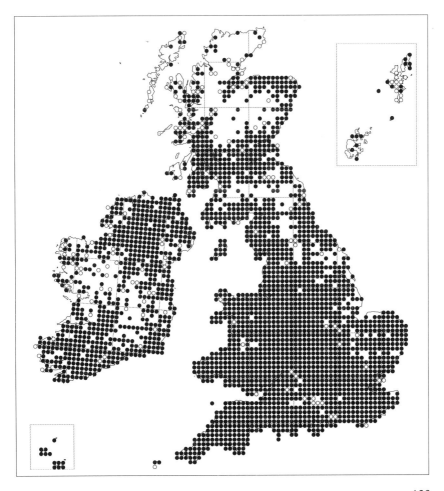

42 Fallopia sachalinensis (F. Schmidt ex Maxim.) Ronse Decr.

Giant Knotweed, Sakhalin Knotweed

Reynoutria sachalinensis (F. Schmidt ex Maxim.) Nakai

Robust perennial, forming a coarse thicket. Stems annual, 200–400 cm tall, often reddish, woodier at the base and stouter and often taller than 41 *F. japonica* var. *japonica*. Leaves 15–40 cm × 10–24 cm (length:width ratio c. 1.5), oblong, usually acute, cordate at the base, coarse, glabrous above but with sparse long flexuous hairs beneath. Flowers hermaphrodite, with exserted fertile anthers, or male-sterile with included empty anthers, greenish, in panicles shorter and denser than those of *F. japonica*, with the main axis not exceeding 15 cm, densely pubescent, shorter than the leaves. Nut 4–5 mm long, trigonous, dark brown, enclosed by the enlarged, conspicuously winged perianth. 2n = 44* (Bailey & Stace 1992). Flowering in August and September.

Native of N. Japan, Sakhalin, S. Kuriles and Ullungdo (off E. Korea); widely naturalised in N.W. and C. Europe. Introduced into British gardens before 1861, but on account of its size it found a place only in very large gardens and was much less widely planted than *F. japonica*. First recorded in the wild in 1896, it is now widely established as an escape by rivers and roadsides and on commons and waste ground throughout Great Britain and Ireland (map, p.126). It is more tolerant of shade than *F. japonica*. Conolly (1977) gives the history of the species in Britain and Ireland and confirms that some plants are gynodioecious (queried by Hooker 1881).

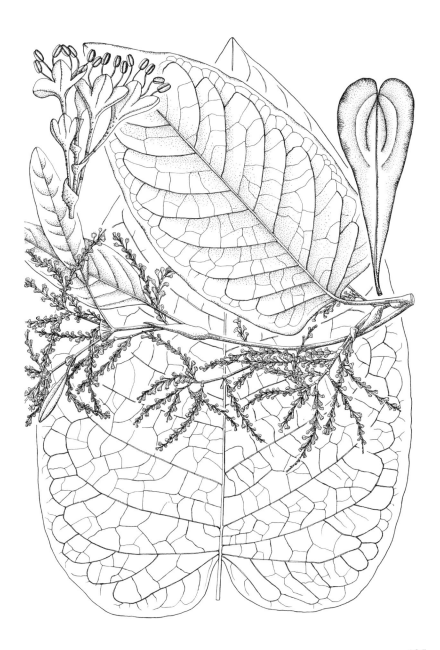

42 Fallopia sachalinensis

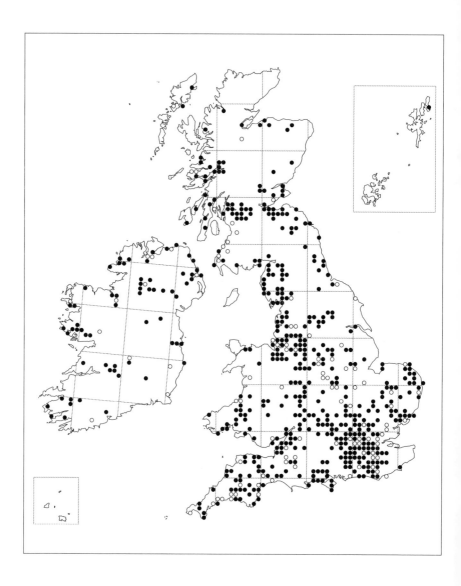

Fallopia × bohemica (p.128) 43

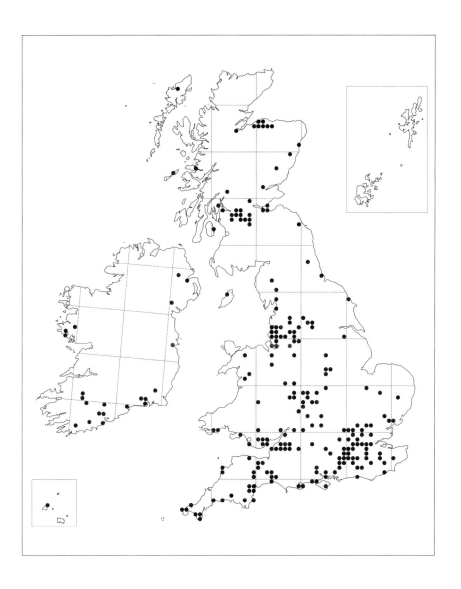

127

43 Fallopia × bohemica (Chrtek & Chrtková) J.P. Bailey

Bohemian Knotweed

41 *F. japonica* (Houtt.) Ronse Decr. × 42 *F. sachalinensis* (F. Schmidt ex Maxim.) Ronse Decr.

Intermediate between *F. japonica* var. *japonica* and *F. sachalinensis* in habit, with stems 250–400 cm tall, somewhat red-spotted. Leaves intermediate in size and shape, up to 23 × 19 cm (length:width ratio 1.1–1.8), acuminate, with the base weakly to moderately cordate; undersides of larger leaves with numerous short, stout hairs (a diagnostic feature). Plants hermaphrodite or male-sterile (gynodioecious). Nut similar to that of *F. sachalinensis*. 2n = 44*, 66*, 88* (Bailey & Stace 1992).

Although not widely recognised by botanists until the 1980s and often confused with *F. sachalinensis*, this hybrid is now known to be widespread in Britain north to Stirling (v.c. 86) and W. & E. Perth (v.cc. 87, 89); in Ireland it is common in Connemara (W. Galway, v.c. H16), beside Lough Neagh, Co. Londonderry (v.c. H40), and probably elsewhere (map, p.127). This hybrid, first collected in 1954, represents a significant element of both the population numbers and genetic variation of *F. japonica* in Great Britain and Ireland, enhancing the genotypic diversity and invasive potential of this complex of taxa (Tiébré *et al.* 2007). Back-crossing with *F. japonica* may occur in Surrey (v.c. 17) and elsewhere (Bailey *et al.* 1996).

Known from the Czech Republic, from which it was first described in 1983, Poland, Belgium and Fennoscandia; probably widespread in Europe wherever the parents occur together, also in North America.

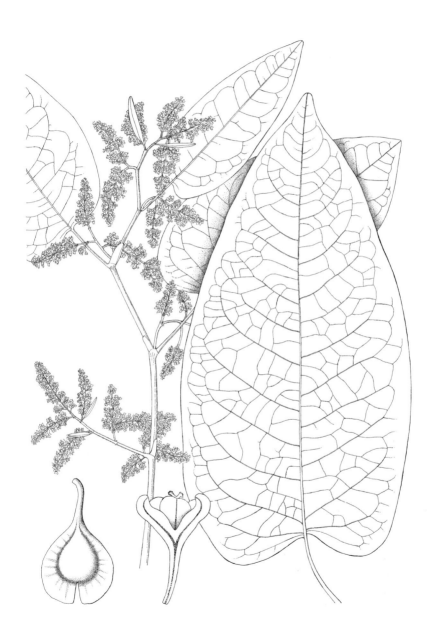

44 Muehlenbeckia complexa Meisn.   *Wire-plant*

A woody, sprawling or climbing shrub, sometimes forming dense curtains of interlacing reddish-brown, wiry stems, which hang down walls and cliffs or scramble amongst other shrubs. Leaves 4–10 mm, on short petioles, dark green above, paler beneath, oblong to suborbicular, entire, coriaceous. Ochreae deciduous. Inflorescence of simple or branched spike-like racemes, axillary or terminal, not more than 2.5 cm. Flowers small, the 5 perianth-segments yellowish-green, fused in the lower third; male with 8 stamens; female with 3 broad, sub-sessile, fimbriate stigmas. Perianth-segments enlarged in fruit, succulent, waxy-white. Nut c. 2 × 1.75 mm, trigonous with blunt angles, black, glossy, partly fused with the perianth. 2n = 20 (Jaretzky 1928). Flowering from July to September.

Native to New Zealand, where it is a variable species; naturalised on Atlantic coasts and islands of Europe. In Britain it is well established as an escape from gardens on low sea-cliffs, rocky and sandy slopes, coastal scrub and hedge-banks in the Channel Isles, Isles of Scilly (where it was first recorded as an escape in 1909) and W. Cornwall (v.c. 1), and probably elsewhere in S.W. England; also in Bournemouth, S. Hants (v.c. 11), and at Bawdsey, E Suffolk (v.c. 25). In Ireland it is recorded from a few places on the coasts of Cos Waterford (v.c. H6), Dublin (v.c. H21) and Down (v.c. H38), but is sometimes rampant in gardens near south-western coasts and likely to spread further.

Muehlenbeckia is a genus of about 20 species in Australasia and temperate South America. Several species are in cultivation (Akeroyd 1989), but only *M. complexa* and *M. axillaris* (Hook. f.) Endl. are at all hardy.

45 Rheum palmatum L.   *Chinese Rhubarb*
(Illustrated on page 133)

Robust, shortly pubescent perennial up to 200 cm tall. Leaves with long, grooved, roughly hairy petiole; lamina suborbicular, deeply palmately lobed, with acute divisions and coarse teeth, somewhat rough above, 3- or 5-veined, roughly hairy on veins beneath and often reddish. Flowers deep red in rather loose panicles with strict, pubescent branches. Fruiting-perianth shortly stalked, 6–10 mm long, the perianth-segments oblong-cordate, winged. 2n = 22 (Jaretzky 1927). Flowering in June and July.

Native of N.E. Asia. Formerly cultivated in Britain as a purgative and increasingly grown as an ornamental; rarely naturalised from gardens, but sometimes

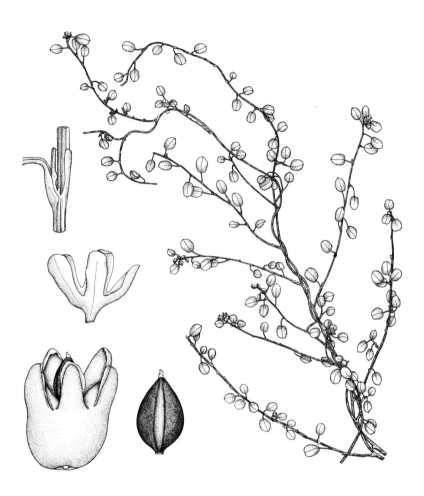

found in rough grassland, on roadsides or beside ponds, with records from S. Somerset (v.c 5), E. Kent (v.c. 15), N. Essex (v.c. 19), Oxon. (v.c. 23), S. Lancs. (v.c. 59) and Westmoreland (v.c. 69).

46 Rheum × hybridum Murray   *Garden Rhubarb*

R. × *cultorum* auct.

Robust perennial 50–200 cm tall. Leaves mostly basal, with a long, stout, fleshy reddish petiole; lamina up to 50 cm long, suborbicular, deeply cordate at the base, entire, 5-veined, dark green, glabrous and shining above, paler and puberulent on the veins beneath. Stem leaves smaller, with a short petiole and more triangular lamina. Flowers whitish in a dense, leafy, fastigiate panicle. Fruiting perianth 7–10 mm long, the perianth-segments ovate-cordate, obtuse, winged. Nut ellipsoid, brown. 2n = 44 (Jaretzky 1928). Flowering in June and July.

Native of S. Siberia; cultivated in Great Britain and Ireland since the 18th century. It is found here and there, especially in northern Britain, as an outcast or survivor of former gardens, around buildings and ruins, on rubbish-dumps, roadsides and waste ground, and on banks of rivers and streams where rhizomes have been washed down.

Cultivated Rhubarb is a complex hybrid that probably involves several Asiatic species, including 45 *R. palmatum* L. (see Foust 1991). A bibliography of Rhubarb and related taxa was provided by Marshall (1988).

47 Rheum officinale Baill.   *Tibetan Rhubarb*

Robust, roughly hairy perennial up to 150 cm tall. Leaves long-petiolate, ovate, deeply 5-lobed, coarsely toothed with irregular acute teeth, 5-veined, pale green; petiole about as long as the lamina. Flowers greenish-cream or pinkish in a dense panicle. Fruiting perianth 7–10 mm, the perianth-segments ovate-triangular, obtuse, winged. 2n = 22 (Jaretzky 1928). Flowering from late May to July.

Native of Tibet and W. China. Introduced to gardens for ornament and as a medicinal plant; occasionally escaping, as at Bolton and Chorley, S. Lancs (v.c. 59), and near Borve in Harris, Outer Hebrides (v.c. 110); there is also a 1937 record from Middleton-in-Teesdale, Co. Durham (v.c. 66).

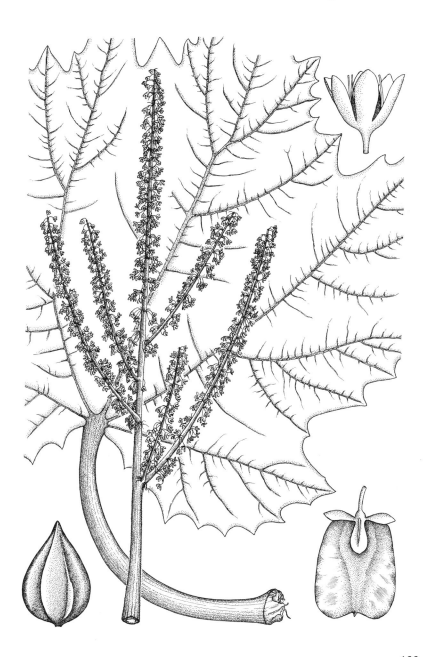

48 Rumex acetosella L. *Sheep's Sorrel*

Erect perennial, tasting of acid, with far-creeping rootstock spreading by shallow roots to form extensive patches; stems 5–50 cm tall but usually about 10–30 cm, slender, flexuous. Leaves very variable, up to 4 cm long, several times longer than broad, linear-lanceolate to broadly lanceolate, usually hastate with 1 pair of spreading lobes and long thin petioles, often somewhat glaucous or flushed with red. Ochreae lacerate, hyaline and often silvery in distal part. Male flowers with 6 obtuse, sepaloid tepals and 6 stamens. Female flowers with 3 sepaloid tepals, enlarging only slightly and closely appressed to the fruit. Nut 0.8–1.5 mm long, trigonous, smooth. Flowering from May to September.

Native. Heathland, non-calcareous sand-dunes, stable river or coastal shingle, and other places with acid or leached, often sandy soils; also on bare serpentine soils or rocky debris and other more neutral soils. Common throughout Britain and much of Ireland, from sea-level to at least 1000 m.

This species, together with 50 *R. acetosa*, is the food-plant of the caterpillars of the Small Copper butterfly (*Lycaena phlaeas*).

A complex species that can be divided, certainly in Britain and Ireland, into the following segregate subspecies on the basis of both cytological and morphological characters. Many botanists have treated these variants as separate species. The adherence or not of the mature valves to the ripe fruit (angiocarpy vs gymnocarpy) can be tested by placing a sample in the palm of one hand and rubbing it hard with the index finger of the other.

1 Valves fused with and not separable from the nut   c. subsp. **pyrenaicus**

1 Valves not fused with and easily separable from the nut   2

  2 Stem erect; inflorescence branching from or above the middle; leaves lanceolate or linear-lanceolate   a. subsp. **acetosella**

  2 Stem procumbent to ascending, with upright branches; inflorescence branching from below the middle; leaves linear, the basal ones up to 10 times as long as broad   b. subsp. **tenuifolius**

a. subsp. acetosella

Flowering stems more or less erect. Leaves hastate, variable in size and outline, the central lobe lanceolate, often broadly lanceolate. Valves readily separable from nut (gymnocarpous). Nut 1.3–1.5 × *c*. 0.8 mm, trigonous, ovoid, glossy, dark brown. $2n = 14, 28, 42$ (den Nijs 1976).

134

Rumex acetosella subsp. acetosella 48a

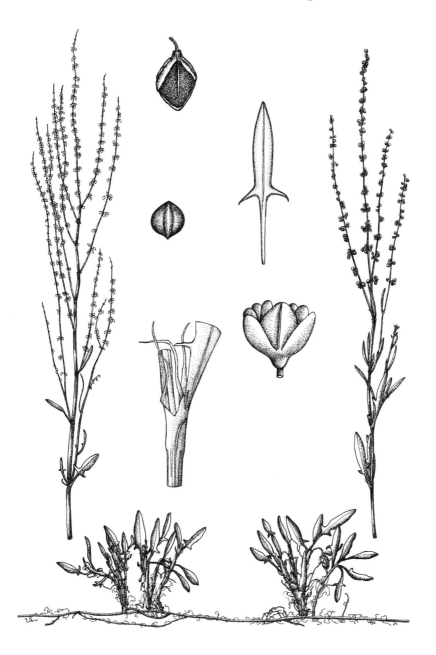

Mainly in northern Britain, but under-recorded. There is a 19th-century Irish specimen from Dingle, Co. Kerry (v.c. H1).

Native of temperate Eurasia; introduced in North America.

b. subsp. **tenuifolius** (Wallr.) O. Schwarz

R. tenuifolius (Wallr.) Á. Löve

Flowering stems ascending, usually bending conspicuously from the base, branched from below the middle. Leaves with central lobe narrowly linear, usually 0.5–1 mm wide and up to 40 times as long, but up to 2 mm wide, the margins often inrolled, the basal lobes small, rarely absent, diverging at right angles. Valves readily separable from the nut (gymnocarpous). Nut 0.9–1.3 mm long, trigonous, ovoid, glossy, light or dark brown. $2n = 28$ (Löve 1940).

Native. Sometimes abundant on dry, humus-deficient, usually sandy soils of heaths and commons. Locally common near southern and eastern coasts of Britain, from Dorset to Sutherland, and on the inland heaths of Surrey, Dorset, Berkshire and the Breckland, with scattered localities elsewhere (map, p. 139). The few Irish records are all coastal, including from heathy sand-dune grassland at Rosslare, Co. Wexford (v.c. H12), where it has not been seen recently.

Widespread in Europe, mainly in the east and north, and in N. Turkey and the Caucasus.

This subspecies is treated by many authors, including Stace (2011), as but a variety (var. *tenuifolius* Wallr.) of subsp. *acetosella*. For this reason, the subspecies has been under-recorded in recent years. In Europe the variation is more continuous and it is not always possible to separate the two variants. Den Nijs (1984), having assessed a body of observational and experimental evidence, has also suggested that this plant merits no more than varietal status. However, in Britain it is for the most part distinct, and a characteristic plant of heathland. Many British botanists have regarded *R. tenuifolius* as a species, a convention recognised here by the compromise use here of subspecific rank.

c. subsp. **pyrenaicus** (Pourr. ex Lapeyr.) Akeroyd

R. angiocarpus auct., non Murb.

Flowering stems more or less erect. Leaves hastate, the central lobe often broadly lanceolate but sometimes narrow, even linear, the basal lobes entire, rarely

136

Rumex acetosella subsp. tenuifolius 48b (A)
subsp. pyrenaicus 48c (B)

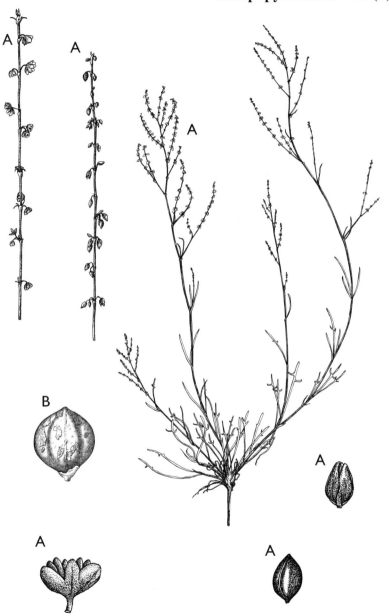

divided, the margins flat. Valves tightly adhering to the nut and difficult to remove (angiocarpous). Nut 1.3–1.5 × *c*. 0.8 mm, trigonous, ovoid, glossy, brown. $2n = 14, 28, 42$ (den Nijs 1974), 56 (Fernandes 1983).

This is the widespread plant and probably the only subspecies present in most of western Britain and in Ireland. It is the most ruderal of the subspecies and can be a persistent weed of sandy fields, flower beds, lawns, grassy banks and tennis courts on light soils. Subsp. *pyrenaicus* occurs throughout W. Europe, including the Iberian peninsula, from where it extends into North Africa. This taxon is now cosmopolitan, even found on oceanic islands, probably widely introduced from the 15th century onwards by the maritime nations of W. Europe via overseas trade and imperial expansion.

The well-known name *R. angiocarpus* Murb. refers to plants, described in 1897 from Bosnia-Hercegovina, with angiocarpous fruits and leaves that have multifid basal lobes. Multifid leaf-lobes, a widespread character in populations of *R. acetosella* in S. & S.E. Europe, are associated with both angiocarpy (subsp. *multifidus* (L.) Arcangeli, incl. *R. angiocarpus* Murb.) and gymnocarpy (subsp. *acetoselloides* (Balansa) den Nijs). All British and Irish plants have leaves with simple, rarely bifid, basal lobes; these plants clearly belong to subsp. *pyrenaicus*, the widespread angiocarpous variant in W. Europe.

Den Nijs (1984) summarised the long, convoluted and sometimes acrimonious debate on the taxonomy of the '*Rumex acetosella* group'. His own observations and experiments have demonstrated that no clear correlation exists between morphology, notably angiospermy vs gymnospermy, and chromosome number. Löve (1983), however, provided a spirited defence of his own view, and that of some other European botanists, that chromosome ploidy level does in fact correlate with morphology.

Rumex acetosella subsp. tenuifolius 48b

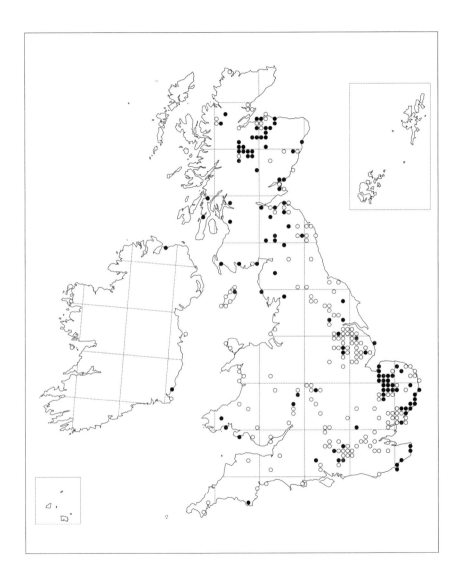

49 Rumex scutatus L.

A glabrous perennial, somewhat woody at the base, with a slender creeping root-stock and straggling stems 30–45 cm long, branched from the base. Leaves 1.5–5.5 cm long, broadly ovate or panduriform, obtuse, hastate, about as broad as long, petiolate, somewhat fleshy, usually glaucous, tasting of acid. Ochreae bifid, silvery and conspicuous. Flowers polygamous in lax, leafless inflorescences, with a few erect branches. Inner tepals *c*. 5 × 5 mm, enlarging in fruit, orbicular, cordate, membranous, pale brown, with attractive reticulated nerves, much exceeding the nut. Nut 3–3.5 mm long, ellipsoid, acute, pale brown, smooth. $2n = 20$ (Kihara & Ono 1926). Flowering from June to August.

Native in Europe, the Caucasus, S.W. Asia and N.W. Africa, where it is widespread on screes, cliffs and rocky ground, and railway ballast and waste places. It has long been grown in gardens in Britain as a culinary herb, the pleasantly acid leaves being used to flavour salads, sauces and soups. Most records of naturalised plants are from old walls of castles and other ancient buildings.

It was established for several centuries in southern Scotland, at Craigmillar Castle, Midlothian (v.c. 83), from where it was lost in 1971 after a wall was repointed, and Aberdour Old Castle, Fife (v.c. 85), where it was said to have been introduced in the late 16th century as a result of the French association with the Stuart monarchs. Many of the other records are from northern England, especially from Settle, Mid-W. Yorks (v.c. 64), where it grows on a road-bridge over the River Ribble. It is rare elsewhere and probably extinct in Ireland, where it was collected over several years at Lisdoonvarna, Co. Clare (v.c. H6).

50 Rumex acetosa L.

Common Sorrel

Acetosa pratensis Miller

a. subsp. acetosa

An erect, nearly glabrous, caespitose perennial, tasting of acid, commonly about 40–70 cm tall, rarely up to 130 cm; rootstock somewhat woody, fibrous with the remains of petiole bases. Basal and lower stem leaves ovate-oblong or -lanceolate, (0.7–)4–7(–15) cm, acute or subacute, hastate with the basal lobes directed more or less parallel to the petiole. Upper stem leaves subsessile, clasping the stem, often narrower and more acute. Ochreae fringed. Plant dioecious. Male flowers in strict panicles with few branches, ceasing to grow when the pollen is shed and, therefore, inconspicuous and often overlooked; female flowers in panicles with ascending branches, continuing to grow as the fruit ripens. Outer tepals reflexed and appressed to pedicels after flowering. Inner tepals in female flowers orbicular-cordate, enlarging to 3–4.5 mm long, membranous, often turning a rich red or purplish-red, especially round the margins. Nut 2–2.5 mm long, glossy, dark brown to blackish-brown. $2n = 14$ (female), $2n = 15$ (male) (Degraeve 1976). Flowering in May and June.

Native. A common species of neutral and slightly acid grassland, also in woodland rides and borders, on coastal and river shingle, and on mountain ledges. Throughout Britain and Ireland, although absent from most 'improved' grasslands and leys.

Widespread in Europe but rarer in much of the south and east, temperate Asia and N.W. Africa; widely introduced in North America.

This species, together with *R. acetosella*, is the food-plant of the caterpillars of the Small Copper (*Lycaena phlaeas*) butterfly.

A variable species, but few variants can be separated satisfactorily. Two coastal subspecies are recognised, but their populations are sometimes intermediate to subsp. *acetosa*.

b. subsp. hibernicus (Rech. fil.) Akeroyd

A small arenicolous plant, papillose-scabrid to puberulent below, with several arcuate, ascending stems, 10–20(–40) cm long, from a thin rootstock. Basal leaves 10–15 × 5–10 mm, with short, acute slightly divergent basal lobes,

Rumex acetosa subsp. acetosa 50a

143

usually with papillate hairs along lower edge, sometimes on both surfaces. Stem leaves few, the upper up to 8 times as long as wide. Panicle dense, with a few short branches or unbranched. Inner tepals up to 4 × 4.5 mm. Flowering in May and June.

Native and apparently endemic. Widespread on sand-dunes and dune grassland (machair) in W. Ireland and N.W. Scotland, rarely elsewhere (map, p.147). Plants so named are apparently not always homogeneous. Some remain constant in cultivation, especially on sandy soil, whilst others change rapidly. This variant was first recorded in the 1950s by M.J.P. Scannell and sent to K.H. Rechinger, who described it at specific rank after seeing it in the field (Rechinger 1961).

Plants of dwarf, congested habit and with papillose-scabrid leaves and stems, from open plant communities on serpentine debris on Unst, Shetland (v.c. 112), from south of Wick of Hagdale and Keen of Hamar, are hardly distinguishable from Irish plants. They somewhat resemble *R. acetosa* var. *serpentinicola* Rune (subsp. *serpentinicola* (Rune) Nordh.) from serpentine habitats in Scandinavia.

Individuals or small populations of subsp. *acetosa* with scabrid-puberulent leaves or petioles belong to var. *hirtulus* Freyn. Small-leaved variants that occur on coastal shingle beaches and in mountain grassland require further study.

c. subsp. **biformis** (Lange) Valdés-Bermejo & Castroviejo

Similar to subsp. *acetosa* but plants much more robust; stems stout, erect; basal and stem leaves often rounded, fleshy; panicle dense, much-branched but the branches mostly simple.

Native to Atlantic coasts of Europe from NW Spain to W Scotland. Local on sea-cliffs in western Britain, for example W. Cornwall (v.c. 1) and Cards (v.c. 46), and western Ireland, for example Kerry Head (v.c. H2) and the Cliffs of Moher, Co. Clare (v.c. H9). As in subsp. *hibernicus*, there is variation within populations, and many plants fall closer to subsp. *acetosa*.

Rumex rugosus Campd., Garden Sorrel, similar to *R. acetosa* and treated by some authors as another subspecies (*R. acetosa* subsp. *ambiguus* (Gren.) Á. Löve), is cultivated for its mildly acid leaves, used to flavour salads, sauces and soups. It is a robust and often luxuriant plant up to 150 cm tall, with large, broad, pale green leaves, dense, repeatedly branched panicles and inner tepals c. 3 mm long. It is unknown in the wild and its origin is obscure, but seed is widely sold and the plant occasionally turns up as an adventive, e.g. at Cheshunt, Herts (v.c.

Rumex acetosa subsp. hibernicus 50b (A) subsp. biformis 50c (B)

20), Felixstowe, E. Suffolk (v.c. 25), and Cambridge, Cambs. (v.c. 29). $2n = 14$ (female), 15 (male) (Degraeve 1976).

Two closely related European species in *Rumex* section *Acetosa* have been recorded in error from Britain:

R. alpestris Jacq. was reported by Druce (1924), as *R. arifolius* All., from mountain cliffs at Lochnagar, Aberdeen (v.c. 92). However, the specimens represent a variant of *R. acetosa* with thinner, broader leaves than usual that have rather spreading auricles, and lax panicles. This species is widespread in Fennoscandia and the mountains of C. & S. Europe (Jalas & Suominen 1979).

R. thyrsiflorus Fingerh. was recorded once from Three Cocks, Hereford. (v.c. 36), but the specimens seen all belong to *R. acetosa*. The species differs from *R. acetosa* by the basal leaves being 3–4 times as long as wide, lanceolate and crisped, with the basal lobes directed outwards, the dense, repeatedly branched panicles, and the valves 2.5–3(–4) mm. A native of dry grassland in C. and E. Europe, flowering about six weeks later than *R. acetosa*, it has spread into W. Europe (Rechinger 1964; Jalas & Suominen 1979), especially along road-verges.

Rumex acetosa subsp. hibernicus 50b

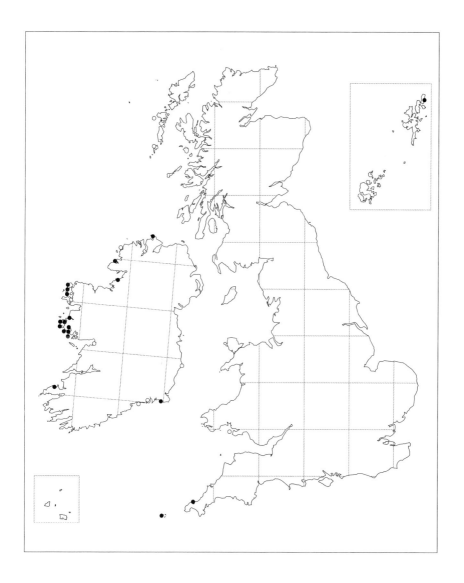

51 Rumex salicifolius T. Lestib. var. triangulivalvis (Danser) J.C. Hickman

Willow-leaved Dock

R. triangulivalvis (Danser) Rech. fil.

A perennial with a creeping rhizome; stems 20–50(–100) cm tall, sometimes decumbent at the base, erect, flexuous, branched. Leaves 8–12(–20) × 1–3(–4) cm, linear-lanceolate, gradually narrowed at both ends, acute, shortly petiolate or subsessile, pale green and papery when dried. Panicle open, with a few simple arcuate-ascending branches; as each panicle passes into fruit a fresh flowering shoot grows up from a lower axil and eventually exceeds the primary panicle; whorls congested above, distant below, the lower subtended by long linear leafy bracts. Valves 3–4 × 2.5–3 mm, triangular, entire or finely denticulate, reticulate, each with an elongate muricate tubercle which occupies only about a quarter of the total width of the valve. Nut 2–2.3 × 1.3–1.8 mm, dark brown, trigonous with acute angles, broadest near the middle, with an acuminate apex. $2n = 20$ (Kihara & Ono 1926, Sarkar 1958). Flowering from June to September.

Native of North America, where it is a widespread and extremely variable ruderal (Sarkar 1958); also as an introduction in N. & C. Europe, mostly brought in with grain. In Britain it has been recorded from widely scattered stations, mostly about docks and railways, and on waste ground and rubbish-tips. It has been most frequently recorded in southern and eastern England, especially in the London area, where it has been known since at least 1909. Seed germinates freely and remains viable for at least five years (Lousley 1944). Usually sporadic in occurrence, *R. salicifolius* persists in at least one locality in S. Essex (v.c. 18). There are other scattered records from northern Britain and two pre-1920 records from Ireland.

Another widespread North American species, **R. altissimus** Alph. Wood, Tall Dock, up to 1m or more, with lower leaves ovate- to oblong-lanceolate, valves 4–6 × 3-4(–5) mm, broadly ovate, entire, only one bearing a tubercle, and nut *c.* 3 mm long, has been recorded as a casual in N. Essex (v.c. 19) and Middx (v.c. 21) (Rechinger 1948). $2n = 20$ (Sarkar 1958). It has not been not seen after 1930.

R. salicifolius, *R. altissimus* and 52 *R. cuneifolius* belong to *Rumex* section *Axillares* Rech. fil., characterised by a lack of basal leaf rosettes and the indeterminate growth of the panicle over an extended flowering period. This section is largely North American, with other species found mostly in South America, South Atlantic islands and the Pacific region (Rechinger 1984, 1990).

52 Rumex cuneifolius Campd.   *Argentine Dock*

R. frutescens Thouars

An arenicolous perennial with a far-creeping woody rhizome, which spreads horizontally at about 35 cm below the sand surface, sending up aerial shoots 15–30 cm tall. Leaves 5–12 × 3–8 cm, obovate, obtuse, often cuneate at the base; margins finely crisped and crenate, all very coarse and leathery. Panicle congested, with a few short simple branches; secondary panicles sometimes arising from lower axils and eventually exceeding the primary one. Valves 4–5 × 2.5–3 mm, ovate-triangular, rather acute, entire, coriaceous, all bearing a large elongate tubercle with a finely punctate surface. Nut 2–2.5 mm, ovoid, broadest near the middle, glossy, brown. $2n = 40$ (Löve 1986), 160 (Ichikawa *et al.* 1971). Flowering in July and August.

Native of temperate South America, where it is widespread on coastal dunes and elsewhere; occasionally adventive in North America and a sporadic casual in W. Europe. In Britain it has been established in dune-slacks on Phillack Towans, W. Cornwall (v.c. 1) since 1921, Braunton Burrows, N. Devon (v.c. 4) since 1929, and Kenfig Burrows, Glamorgan (v.c. 41) since 1934. There is also a 1913 record from Wallasey sandhills, Cheshire (v.c. 58). Seed may have been conveyed to these places by sea.

R. cuneifolius was also temporarily established at docks: on the Bristol Channel at Sharpness and Avonmouth, W. Gloucester (v.c. 34), at Portishead, N. Somerset (v.c. 6), and Cardiff, Glamorgan (v.c. 41), and in Scotland at Gartcosh, Glasgow, Lanark (v.c. 77), for 10 years and Leith, Mid-Lothian (v.c. 83), for 20 years. It has also occurred here and there as a wool alien.

Rumex cuneifolius was formerly a synonym of *R. frutescens* Thouars, a name familiar to British botanists but which now applies to a similar species in section *Axillares* endemic to the islands of Tristan da Cunha and Gough in the South Atlantic (Rechinger 1990).

Hybrids

R. × wrightii Lousley (52 *R. cuneifolius* × 64 *R. conglomeratus*)

A rhizomatous perennial 30–40 cm tall. Leaves coriaceous like those of *R. cuneifolius*, but truncate or subcordate at the base. Panicles with long branches in remote whorls subtended by bracts below. Valves up to 5 mm long, lingulate, with 3 tubercles. Recorded from sand-dune grassland at Phillack Towans, W. Cornwall (v.c. 1) since 1982 (Holyoak 1995) and dune-slacks at Braunton Burrows, N. Devon (v.c. 4) (Lousley 1953a). Unknown outside Britain.

150

R. × **cornubiensis** D.T. Holyoak (52 *R. cuneifolius* × 74 *R. obtusifolius*)

A rhizomatous perennial more or less intermediate between the parents but distinctly taller than *R. cuneifolius*. Leaves more coriaceous than those of *R. obtusifolius*, truncate to weakly cordate at the base, more or less crenulate, the midrib and main veins papillose-scabrid beneath. Valves up to 6 mm long, shortly toothed, with 3 tubercles; often withering, the fruits being almost completely infertile. Sand-dunes at Phillack Towans, W. Cornwall (v.c. 1) (Holyoak 1995); unknown outside Britain.

53 Rumex pseudoalpinus Höfft   *Monk's Rhubarb*

R. alpinus sensu L. (1759), non L. (1753)

A stout rhizomatous perennial with erect stems 60–70(–100) cm tall, extensively creeping to form large patches. Plant light green; stems and petioles sometimes tinged with red. Basal leaves 20–40 × 20–35 cm, orbicular, cordate, with slightly undulate margins; petiole longer than the lamina. Panicle fusiform in outline, dense, with fasciculate branches, leafy in the lower part, leafless above, rarely interrupted; pedicels filiform, up to 3 times as long as the tepals, deflexed in fruit. Valves 4.5–5 × 3.5–5 mm, ovate, subobtuse, truncate at the base, entire or finely denticulate towards the base, membranous, without tubercles. Nut *c.* 3 mm, trigonous, light olive-brown. $2n = 20$ (Kihara & Ono 1926). Flowering from June to August.

Native to the mountains of C. and S. Europe, N. Turkey and the Caucasus; introduced to Britain and established near farms, by streams and on roadsides, mainly in hilly districts from Stafford (v.c. 39) and Derby (v.c. 57) northwards; locally common in Scotland (map, p.155). It has been recorded occasionally from further south, for example Surrey (v.c. 17). A 1957 record from Ireland, from Black Mountain, Co. Antrim (v.c. H39) needs confirmation.

In its native range *R. pseudoalpinus* is a member of nitrate- and phosphate-enriched 'Lägerflur' communities manured by livestock, around mountain villages, farms, barns and sheepfolds, together with other nitrophilous plants such as *Chenopodium bonus-henricus* and *Urtica dioica*.

This species was formerly used widely for human and veterinary medicine and is still valued in the Alps and Carpathians for the treatment of sore places on the udders of cows. It is probable that it was for uses such as this that *R. pseudoalpinus* was dispersed among northern farming communities in Britain, rather than, as is often suggested, because the leaves were used for wrapping butter.

Hybrids

No hybrids have been reported from Britain. The hybrid with 74 *R. obtusifolius* (*R. × mezei* Hausskn.) is known from the Alps, Tatra Mountains and Romanian Carpathians, and also from Finland.

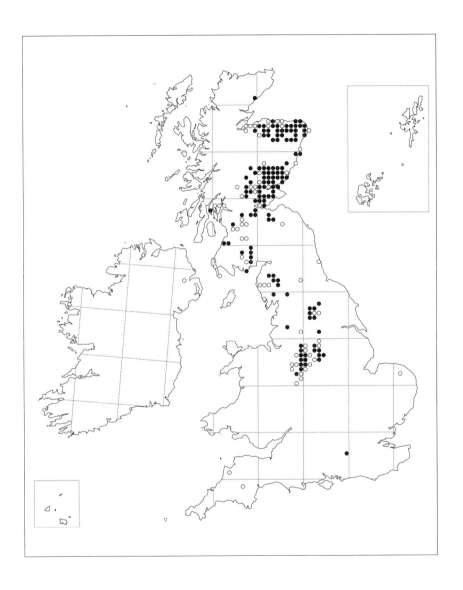

54 Rumex aquaticus L. Scottish Dock

A stout perennial with erect stems 100–200 cm tall. Basal leaves 20–45 × 10–22 cm, triangular, acute or subacute, undulate, deeply cordate at the base, bluish-grey beneath; petiole variable in length, often equalling the lamina. Stem leaves 10–25 × 4–13 cm, increasingly narrower above. Panicle dense, with many ascending branches and a few leafy bracts below; pedicels filiform, up to twice the length of the valves. Valves (4–)6–8 × 4–6 mm, ovate-triangular, more or less acute, usually longer than wide, truncate at the base, entire, without tubercles. Nut 2.5–4 mm long, trigonous, light brown. $2n = 120$ (Löve 1986), 140 (Ichikawa *et al.* 1971), *c.* 200 (Jaretzky 1928). Flowering in July and August.

Native to Arctic Eurasia and C. Europe, extending southwards to C. France and the Balkan peninsula; also North America. In Britain it is restricted to Loch Lomond in central Scotland, on silty and gravelly lochsides, beside a river and ditches, and at the edge of Alder carr. It occurs at the south-eastern corner of the loch in Stirling (v.c. 86) and Dunbarton (v.c. 99) and on its west bank in Dunbarton (v.c. 99). Absent from Ireland.

Before 1939 *R. aquaticus* in Britain was confused with 55 *R. longifolius* and to a lesser extent with 58 *R. hydrolapathum*. The occurrence of the species in Scotland was established by Lousley (1939a). Idle (1968), in a morphological study of *R. aquaticus*, pointed out that it could be distinguished from *R. longifolius* by its general size and height, the size and shape of the basal leaves and fruits, and the structure of the inflorescence and fruiting pedicel.

Although *R. aquaticus* has such a restricted range in Britain (map, p.158), it is widespread in wet places in N., C. & E. Europe. Its limited distribution in Scotland makes it potentially vulnerable to any threat to its habitat (IUCN global category: not threatened; UK category: Vulnerable).

Hybrids

Three hybrids are known in Scotland. Elsewhere in N. Europe hybrids with 55 *R. longifolius* (*R.* × *armoraciifolius* Neuman) and 58 *R. hydrolapathum* (*R.* × *heterophyllus* Schultz) are frequent; the hybrid with 64 *R. conglomeratus* (*R.* × *ambigens* Hausskn.) is also known in Europe and is likely to occur in Scotland.

R. × conspersus Hartm. (54 *R. aquaticus* × 61 *R. crispus*)

Similar in habit to *R. aquaticus* but with leaves crisped and the panicle slighty more lax and leafy. Variably fertile; valves 4–6 × 3.5–4 mm, often with one very small, elongate tubercle. Marshes on the Stirling (v.c. 86) and Dunbarton (v.c. 99) shores of Loch Lomond, growing with both parents; first recorded in

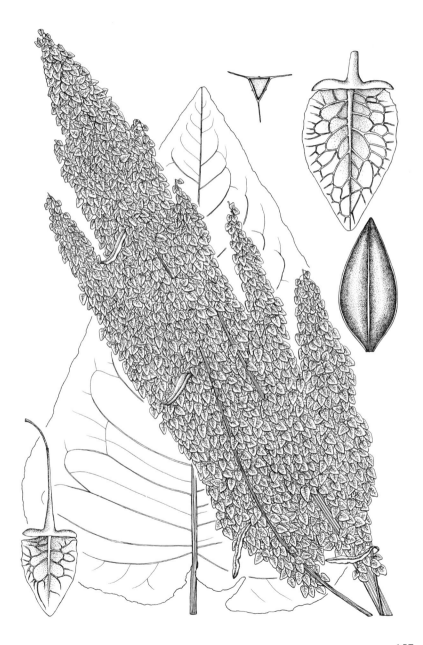

Scotland in 1976 but not confirmed until 1981 (Mitchell 1982). Recorded in Fennoscandia, Austria and Germany; in Sweden there is evidence of introgressive hybridisation between the parents.

R. × **dumulosus** Hausskn. (54 *R. aquaticus* × 65 *R. sanguineus*)

Intermediate in height and habit between the parents, with the triangular lower leaves of *R. aquaticus* and the lax panicle of *R. sanguineus*. Fertility low, but the few valves developed are narrowly oblong like those of *R. sanguineus*, with a single small globose tubercle. Known only from a winter-flooded clearing in damp oak woodland at Curraghlas Wood, Stirlings (v.c. 86); first reported in 1989 (Mitchell 1990).

R. × **platyphyllos** Aresch. (54 *R. aquaticus* × 74 *R. obtusifolius*)

Similar to *R. aquaticus* in habit but the basal and lower cauline leaves more ovate and less triangular in outline, subobtuse. Panicle lax, rather leafy below, with irregularly maturing, more or less acute valves 5–7 × 3.5–6 mm, distinctly toothed near the base and reddish in colour. There is evidence of introgressive hybridisation (Hull & Nicholl 1982), although fertility is often low. Restricted to a small area between Balmaha and Gartocharn on or near the shore of Loch Lomond in Stirling (v.c. 86) and Dumbarton (v.c. 99), where it is common. Recorded from several countries in N. and E. Europe.

54 Rumex aquaticus

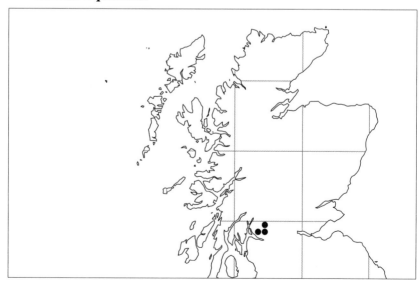

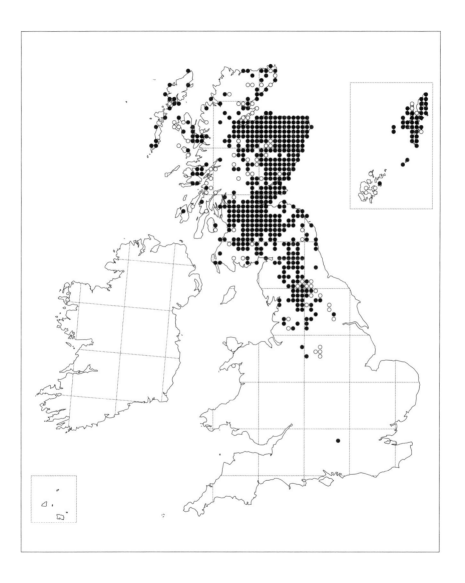

159

55 **Rumex longifolius** DC. *Northern Dock*

Rumex domesticus Hartm.

A stout perennial with erect stems 60–120 cm. Basal leaves broadly lanceolate, up to 60 cm long, 3–4 times as long as wide, the base truncate to cordate or rarely attenuate, the apex usually subacute, the margin crenulate, undulate, of a thin texture. Stem leaves similar but gradually merging into the narrowly lanceolate bracts. Panicle very dense, compact, fusiform, with long leafy lanceolate bracts. Valves 4.5–5.5 × 5.0–6.5 mm, reniform, slightly cordate at the base, obtuse to rounded at the apex, entire, without tubercles. Nut 3–4 mm, mid-brown, trigonous with acute angles, broadest near the middle, narrowed suddenly to the base, and gradually to the apex. $2n = 40$ (Hamet-Ahti & Virrankovsky 1970), $2n = 40, 60$ (Jalas & Lindholm 1975), $2n = 60*$ (Montgomery *et al.* 1997). Flowering in July and August.

Native. On open, disturbed ground by rivers and lakes, in pastures and around arable fields, and about farms. It is a plant of northern Britain (map, p.159), common in southern and eastern Scotland, less so on western coasts and islands, extending into northern England south to Staffs (v.c. 39) and Cheshire (v.c. 58); also in Orkney (v.c. 111) and Shetland (v.c. 112); absent from Wales and Ireland.

Native in Europe south to Denmark, and especially common in Scandinavia, with outlying populations in the Pyrenees and Massif Central; also in Siberia and introduced in North America. Jalas & Lindholm (1975) studied morphological and chromosomal variation in plants from Scandinavia and the Pyrenees.

Hybrids

R. × propinquus J. Aresch. (55 *R. longifolius* × 61 *R. crispus*)

Similar to *R. longifolius* in habit, but the panicle less dense, and sometimes little branched; stem leaves usually broadly lanceolate; valves variable in size, slightly toothed, more triangular and less rounded than in *R. longifolius*, usually only one bearing a tubercle. Fertility is variable. Locally common in Scotland wherever the parents grow together. Widespread and common in Fennoscandia.

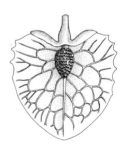

R. × hybridus Kindb. (*R. × arnottii* Druce) (55 *R. longifolius* × 74 *R. obtusifolius*)

A variable, tall and rather untidy plant. Leaves similar to those of *R. longifolius*

160

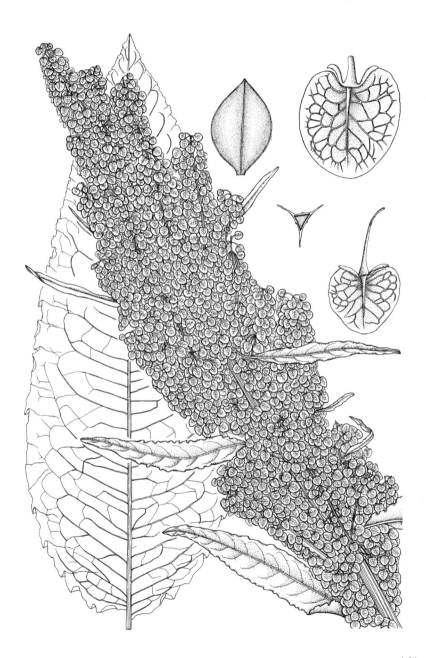

but narrower, usually broadly cordate at the base and papillose-scabrid beneath. Panicle much laxer than that of *R. longifolius*, often reddish. Valves, many of which mature, of about the same size as those of *R. longifolius* but more triangular and less obtuse, with occasional tubercles and usually a few teeth at the base. Fertility low. Locally common in ruderal habitats in Scotland, especially in the eastern half, and northern England, wherever the parents grow together. Widespread in Fennoscandia; also in the Pyrenees.

Hybrids with several other species are known in Fennoscandia (Karlsson 2000).

56 Rumex pseudonatronatus Borbás

Perennial with a fusiform root; stems usually solitary, 50–120 cm tall, erect, unbranched and often suffused with purple. Lower leaves 15–30 × 1–3 cm, linear-lanceolate, narrowed at both ends, subacute, crisped, 8–15 times as long as broad; petiole up to half as long as the lamina. Upper leaves smaller, with a shorter petiole. Panicle long, dense, thyrsoid, interrupted below, more or less leafless. Valves 3.5–5 × 3–5 mm, broadly triangular-ovate, subcordate at the base, obtuse, entire, all without tubercles (or with a single inconspicuous tubercle). Nut 2–3 mm, pale brown. $2n = 40$ (Mulligan 1957).

Native of stony and gravelly riversides and seashores in Siberia and arctic and E. Europe west to Sweden and Austria. Introduced to Britain as a rare casual and apparently recorded only from Hackney Marshes, Middlesex (v.c. 21), in 1910 (Kent 1975) and Sharpness Docks, W. Gloucs (v.c. 34) in 1981.

Records of *R. pseudonatronatus* from Cheshire (v.c. 58) (Lousley 1967) are erroneous and refer to 55 *R. longifolius* (Lousley 1969). *R. pseudonatronatus* can be distinguished from that species by the broader leaves, almost leafless panicle and somewhat smaller valves in fruit. In N. Europe it forms hybrids with several other *Rumex* species.

57 Rumex confertus Willd.   *Russian Dock*

A robust, scabrid-pubescent perennial with stems up to 80 cm tall, similar in habit to 53 *R. pseudoalpinus*. Basal leaves with lamina 20–27 × 15–24 cm, broadly cordate-ovate, obtuse, broadest near the base, hairy below when young and scabrid along veins; petiole often longer than the lamina. Stem leaves narrower, usually somewhat triangular in outline. Panicle large, dense, the branches arcuate at the base. Valves large, 6–8(–11) mm long, cordate, usually broader than long, rounded to slightly apiculate, entire or with a few small, irregular obtuse teeth or crenulate, one with a small tubercle. Nut 3–3.5 × 2–2.5 mm, brown. $2n = 40$ (Löve & Löve 1961), 100 (Löve & Löve 1961). Flowering from late June to August.

Native from Siberia to E. Europe, extending south to the Crimea and Caucasus; during the 19th and 20th centuries it spread across Poland and further west (Trzcinska-Tacik 1963). In Britain known only from two colonies, one of them previously referred to 56 *R. pseudoalpinus*, near Maidstone, E. Kent (v.c. 15); formerly known from Marston, Oxon (v.c. 23), where it persisted from 1918 till *c*. 1930, probably originating from Oxford Botanic Garden (in 1918 named in error as *R. giganteus,* see p. 226), and also as a persistent casual at Old Coulsdon, Surrey (v.c. 17), between 1942 and 1960.

The basal leaves of *R. confertus* closely resemble those of 53 *R. pseudoalpinus* but are slightly less orbicular and more or less obtuse. The stem leaves, however, can be confused with those of 60 *R. patientia*, from which they may be distinguished by their more deeply cordate bases and the scabrid veins beneath when young. The much broader and often larger valves of *R. confertus* should also separate it from *R. patientia*.

Hybrids

Two hybrids have been recorded in Britain; several others occur in E. and N.E. Europe.

R. × **skofitzii** Blocki (57 *R. confertus* × 61 *R. crispus*)

A tall, robust plant resembling *R. confertus* but with narrower cauline leaves, with crisped margins, a lax, narrow panicle, with rather strict branches, and lanceolate, crisped bracts. Valves maturing irregularly, with a less rounded apex than in *R. confertus*; fertility variable. Recorded in 1954–1955 at Old Coulsdon, Surrey (v.c. 17), where *R. confertus* was then naturalised. Known from Finland and several contries in E. Europe.

R. × **borbasii** Blocki (57 *R. confertus* × 74 *R. obtusifolius*)

Intermediate in habit and height between the parents, the basal and lower cauline leaves similar to those of *R. confertus* but thinner in texture. Panicle lax, the valves maturing irregularly, subacute, with sharply toothed margins. Recorded in 1954 at Old Coulsdon, Surrey (v.c. 17), where *R. confertus* was then naturalised. Known from E. Europe, with evidence for introgressive hybridisation in Finland.

58 Rumex hydrolapathum Hudson   *Water Dock*

A robust, erect perennial with several stems 100–200 cm from the stout, black rootstock, often forming a conspicuous waterside tussock. Basal leaves up to 120 cm long, sometimes longer, 4–5 times as long as wide, broadly lanceolate, flat, coriaceous, acute or acuminate, cuneate or subcordate at the base, dull green; lateral veins more or less at 90° to the midrib; margins entire. Stem leaves and bracts narrower, narrowed equally to both ends, sometimes finely crenulate. Panicle much branched, leafy at the base, with branches ascending, whorls rather crowded; pedicels stiff, *c*. 1.5 times as long as the tepals. Valves 5–7 × 3–4 mm, triangular, truncate or subcuneate at the base, usually with a few inconspicuous teeth near the base, each with a prominent, elongate tubercle extending for over half its length. Nut 3–5 mm long, trigonous with sharp angles, chestnut-brown. $2n = c$. 200 (Kihara & Ono 1926), 130* (Montgomery *et al.* 1997). Flowering from July to September.

Native. By rivers, streams, canals, lakes and ponds, and in marshy places and ditches, usually as an emergent. Generally distributed in England, Wales and Ireland (except much of the west), but much rarer in Scotland, where it is mainly coastal; especially common in southern England (map, p.167). It is sometimes planted for ornament, so its status can be doubtful, especially in the northern part of its range.

Native of Europe, where it extends from Finland and Scotland south to the Pyrenees, S. Italy and N.C. Greece, and a few stations in Turkey and the Caucasus.

This species is the food-plant of the Large Copper butterfly (*Lycaena dispar*), now extinct in Britain.

Hybrids

R. × schreberi Hausskn. (58 *R. hydrolapathum* × 61 *R. crispus*)

A tall, vigorous plant, sometimes exceeding *R. hydrolapathum* in height, with several stems. Basal leaves large, broad, lanceolate, with wavy margins; cauline leaves rather narrow. Valves variable, sometimes similar to those of *R. hydrolapathum* but smaller, at other times similar to those of *R. crispus*, but always with 3 elongate, often reddish tubercles. Fertility low. In Britain it has been recorded from coastal cliffs in Dorset (v.c. 9), marsh ditches in S. Hants (v.c. 11), E. Suffolk (v.c. 25), Notts (v.c. 56) and N.W. Yorks (v.c. 64); in Ireland only from Downpatrick Marshes, Co. Down (v.c.38) (Praeger 1942) and Downhill, Co. Derry (v.c. 40), with an unconfirmed record from Co. Cork. A rare hybrid reported from few European countries.

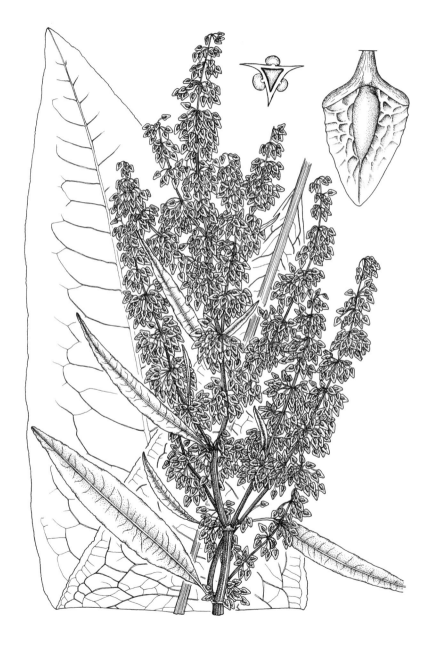

R. × **digeneus** G. Beck (58 *R. hydrolapathum* × 64 *R. conglomeratus*)

A tall plant over 100 cm tall, with large lanceolate leaves, the panicle rather dense, with ascending branches resembling those of *R. hydrolapathum*. Only a few valves maturing, these mostly lingulate, but some broader with large elongate tubercles and closer to those of *R. hydrolapathum*. Rare in Britain by dykes and canals or in adjacent damp ground, recorded from N. Somerset (v.c. 6), E. Sussex (v.c. 14), E. Kent (v.c. 15), Monts (v.c. 47), and Cheshire (v.c. 58); there are no Irish records. A rare hybrid, reported only from Britain, Germany and Hungary.

R. × **lingulatus** Jungner (*R.* × *weberi* Fischer-Benzon) (58 *R. hydrolapathum* × 74 *R. obtusifolius*)

A robust plant similar to *R. hydrolapathum* in size and habit but often taller, with more open panicles and irregular production of fruits. Basal and cauline leaves almost as large as those of *R. hydrolapathum* but tending to be broader, cordate or subcordate at the base, and of thinner texture. Valves similar to those of *R. hydrolapathum* but usually narrower, with a few short but distinct teeth near the base. Fertility low; experiments to germinate the nuts have proved unsuccessful. Local in ditches and dykes, peat cuttings and wet meadows and pastures, mostly in southern England from E. Cornwall (v.c. 2) and S.Somerset (v.c. 5) to E. Sussex (v.c. 14) and E. Norfolk (v.c. 27), extending north to Hereford (v.c. 36), Salop (v.c. 40) and N.E. Yorks (v.c. 62); in Ireland recorded from E. Cork (v.c. H5), Co. Waterford (v.c. H6) and Co. Derry (v.c. H40). A rare hybrid, known also from France, the Netherlands, Denmark and Sweden.

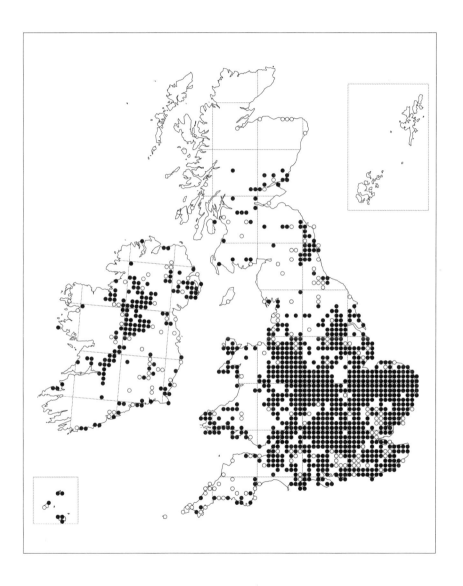

59 Rumex cristatus DC.

Greek Dock

R. graecus Boiss. & Heldr.

A robust perennial with stems 100–200 cm tall, thick, erect, branched above. Basal leaves 20–35 × 8–15 cm, leathery, broadly lanceolate to ovate-lanceolate, cordate at the base, acute or subacute at the apex, with undulate margins and veins (in the middle of the leaf) making an angle of 60–90° with the midrib; petiole up to half the length of the lamina. Stem leaves sessile, ovate or ovate-lanceolate. Panicles large, rather dense, 45–75 cm long, the ascending branches making an angle of about 45° with the axis, with many secondary branches. Valves 5–8 × 5–9 mm, cordate, reddish-brown at maturity, the margins with many subregular to irregular, usually acute teeth up to 1 mm long; one valve with a suborbicular or ovate tubercle 2–3 mm, sometimes the other valves with smaller inconspicuous tubercles. Nut 3–3.5 × *c*. 2 mm, trigonous, dull but with glossy acute angles, dark brown. $2n = 80$ (Ichikawa *et al.* 1971). Flowering in June and July.

Native of the Balkan peninsula, the Aegean region, Cyprus and Sicily, and naturalised in parts of France, Italy and Spain; also locally naturalised in North America. Long introduced and thoroughly established on waste ground, rubbish tips and river banks in Britain (map, p.171), especially the London area and Thames estuary in W. Kent (v.c. 16), S. Essex (v.c. 18) and Herts (v.c. 20); also about the Bristol Channel, from Cardiff, Glamorgan (v.c. 41) and Minehead, S. Somerset (v.c. 5), where it occurs on sand-dunes as well as the usual ruderal habitats, and from Reading, Berks (v.c. 22) and other sites in the Thames Valley.

Hybrids

R. × xenogenus Rech. fil. (59 *R. cristatus* × 60 *R. patientia* L.)

Intermediate between the parent species and with panicles superficially similar in general appearance to 62 R. × *pratensis* (61 *R. crispus* × 74 *R. obtusifolius*), but with low fertility, the valves falling early to produce a ragged appearance. Leaf-veins in the middle of the leaf making an angle of 60–75° with the midrib. Valves rather irregular, slightly smaller than those of *R. cristatus*, the margins with uneven teeth up to 0.5 mm long; usually one ovoid tubercle 1–2.3 mm, sometimes two other very small tubercles present. Known only from Rainham Marshes, S. Essex (v.c. 18) (Kitchener 2002, with an illustration). Rare in Europe, reported only from Britain and Austria (Vienna) (Rechinger 1948). The British plants appear to have *R. patientia* subsp. *orientalis* as one parent, while the original description was based on a hybrid involving subsp. *patientia* (Kitchener 2002).

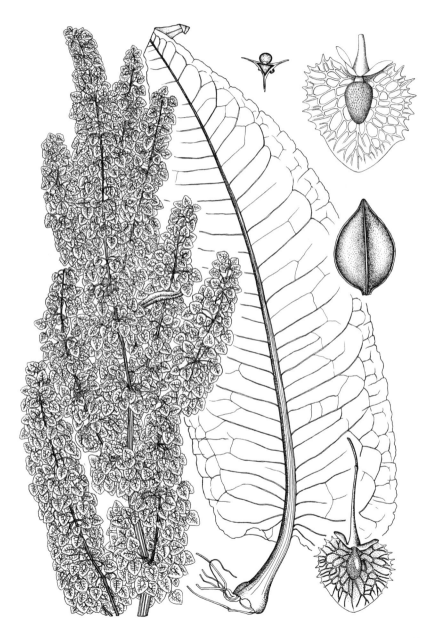

R. × **dimidiatus** Hausskn. (59 *R. cristatus* × 61 *R. crispus*)

A tall, robust plant resembling *R. cristatus*, but with the valves usually smaller, with a single ovoid tubercle and small irregular teeth. Cauline leaves narrow, with crisped margins. Mostly infertile. In Britain recorded only from about the Thames estuary in W. Kent (v.c. 16) and S. Essex (v.c. 18), from Minehead, S. Somerset (v.c. 5), and suburban London in Herts (v.c. 20) and Middlesex (v.c. 21). Also known from Greece and Sicily.

R. × **lousleyi** D.H. Kent (59 *R. cristatus* × 74 *R. obtusifolius*)

Plant up to 200 cm tall, superficially resembling *R. cristatus*. Basal leaves thick, broadly lanceolate, subacute, cordate at the base, with the margins undulate. Cauline leaves smaller, narrowly lanceolate, toothed. Panicle large, 45–60 cm long, with ascending smaller branches; whorls crowded, subtended by bracts below. Valves triangular, reticulate, like those of *R. cristatus* but with a larger ovoid tubercle and irregular, acute teeth. Recorded from Minehead, S. Somerset (v.c. 5), Weston-super-Mare, N. Somerset (v.c. 6), in and around London in Kent (v.cc. 15 & 16), S. Essex (v.c. 18) (Jermyn 1975, first record) and Middlesex (v.c. 21) (Kent 1977, locus classicus), and from Wales, in Monmouth (v.c. 35) and Glamorgan (v.c. 41). Reported only from Britain and (probably involving another *R. obtusifolius* subspecies) Greece.

R. × **akeroydii** Rumsey (59 *R. cristatus* × 77 *R. palustris*)

In habit like a large plant of *R. palustris*, with long arcuate-ascending branches, but much more robust. Infructescence dark reddish-brown, with remote whorls of fruits. Valves superficially similar to those of the alien 75 *R. dentatus*, with lingulate apex and long, irregular, acute teeth. Known only from Rainham Marshes, S. Essex (v.c. 18) (Rumsey 1999, with an illustration; Kitchener 2002), and unknown outside Britain. The native ranges of the parent species overlap only in N. and C. Greece, where *R. palustris* is at the edge of its range.

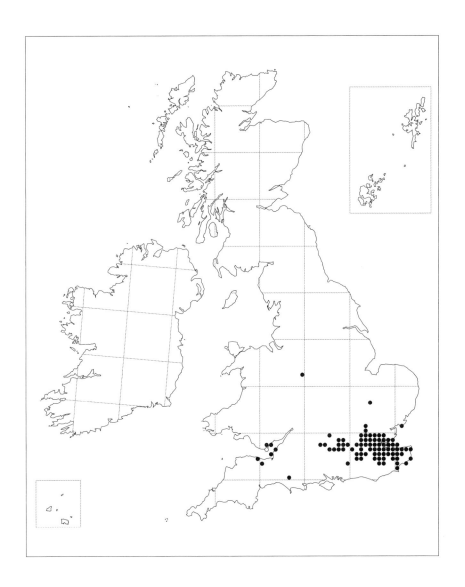

60 Rumex patientia L.

Patience Dock

A robust glabrous perennial with stems 80–200 cm tall, thick, erect, branched above. Basal leaves 35–45 × 10–13 cm, 3-4 times as long as wide, leathery, ovate- or oblong-lanceolate, truncate or subcuneate at the base, rarely subcordate, subacute at the apex, pale green, the veins (in the middle of the leaf) making an angle of 45–60° with the midrib; petiole 25–30 cm, usually longer than the lamina. Stem leaves narrower and more acute, the margins undulate. Panicles 45–50 cm, with short ascending branches, somewhat leafy below. Valves 5–8 × 5–10 mm, broadly ovate or suborbicular-cordate, cordate at the base, blunt or slightly pointed at the apex, the margins entire, crenulate or minutely denticulate, one with a tubercle, the other two sometimes with smaller tubercles. Nut 3–3.5 × 2 mm, trigonous, dull but with glossy acute edges, light brown. $2n = 60$ (Kihara & Ono 1926). Flowering in May and June.

Native of E. and S.E. Europe, and S.W. Asia, and introduced in North America; formerly cultivated as a leaf vegetable in C. and W. Europe and widely naturalized there. In Britain it was never a popular vegetable, and is likely to have been introduced by commerce, hence its occurrence at or near docks, wharves, riverbanks, gasworks and breweries. It has long been established on waste ground and riverbanks in urban areas of England, especially around London, Bristol, Burton-on-Trent and Manchester. There are no records from Ireland.

Two subspecies are present in Britain, but they are not always easy to distinguish.

a. subsp. patientia

Stem purplish- or reddish-brown. Valves 5–7 × 5–8 mm, blunt, one with an indistinct globose or ovoid tubercle *c*. 1.5 mm long. $2n = 60$ (Kihara & Ono 1926).

Native of S.E. Europe, from Slovakia to northern Greece, and S.W. Asia. Established on waste ground about London, particularly by the Thames and elsewhere, but much less frequent than subsp. *orientalis*.

b. subsp. orientalis Danser

Stem green, often up to 2 m tall. Leaf-base more deeply cordate. Valves (5–)6–9 × 8–10 mm, one with a conspicuous ovoid tubercle 2–3 mm long, the other two sometimes with smaller tubercles. $2n = 60$ (Kihara & Ono 1926).

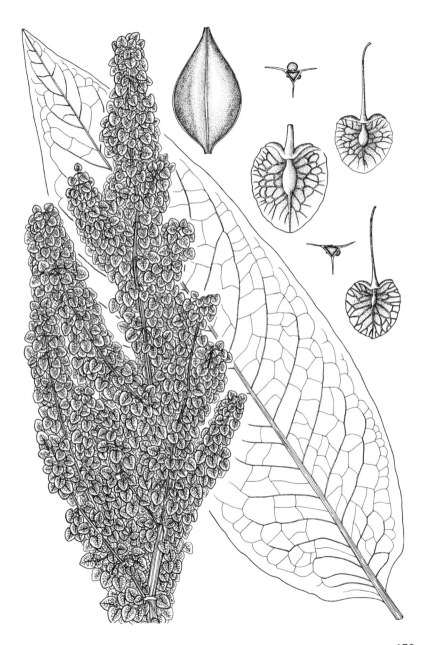

Hybrids

R. × **confusus** Simonkai (60 *R. patientia* × 61 *R. crispus*)

A tall, robust plant closely resembling *R. patientia* but with much smaller valves similar to those of *R. crispus* and narrower, usually crisped upper cauline leaves. It has been recorded from E. Kent (v.c. 15), S. Essex (v.c. 18), Middlesex (v.c. 21) and W. Gloucs (v.c. 34). In Europe it is widespread within the distribution of *R. patientia*, especially in Austria, Hungary and Romania (Transylvania).

R. × **erubescens** Simonkai (60 *R. patientia* × 74 *R. obtusifolius*)

A tall, robust plant somewhat resembling *R. patientia* but with leaves much broader in relation to their length and papillose-scabrid on the veins beneath, branches longer and more spreading, and smaller, reddish valves with short teeth on each margin. Found with *R. patientia* about the River Thames in W. Kent (v.c. 16), Surrey (v.c. 17), S. Essex (v.c. 18) and Middlesex (v.c. 21); also recorded from Sharpness Docks, W. Gloucs (v.c. 34), and Burton-on-Trent, Staffs (v.c. 39). Known in C. and S.E. Europe within the range of *R. patientia*.

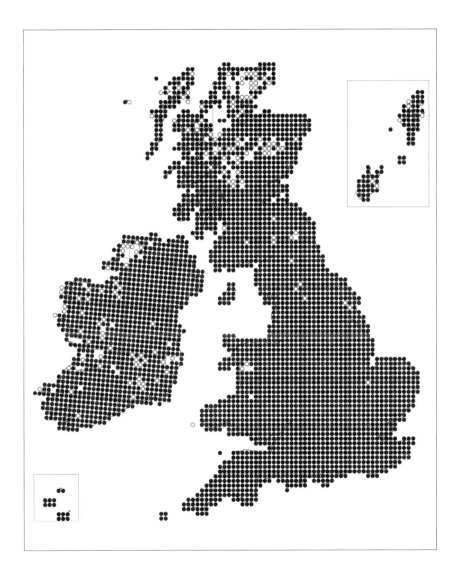

61 Rumex crispus L. *Curled Dock*

An erect, leafy annual, biennial or short-lived perennial, 40–150(–250) cm tall, arising from a stout taproot. Basal leaves 20–40 × 4–6(–10) cm, 4–5 times as long as wide, lanceolate, cuneate, subcordate or sometimes truncate at the base, acute at the apex, slightly fleshy, dull green, the margins usually strongly crisped, crenate; petiole long, often nearly equalling or exceeding the lamina. Stem leaves up to 25 cm, narrowly lanceolate, usually truncate at the base and acuminate at the apex, with the margins crenate, less crisped. Panicle lax or dense, subsimple or with short suberect branches making an angle of not more than 30° with the main stem, the flowers in crowded whorls of 10–30; pedicels 2–10 mm, filiform. Valves 3–5.5(–6.5) × 3–4.5(–6) mm, orbicular-deltate, cordate or subcordate at the base, obtuse or rarely subacute, finely reticulate, entire or occasionally crenate or with a few minute teeth (sometimes only 1 or 2) with 3, sometimes 1 (f. *unicallosus* (Peterm.) Lousley), well-developed smooth, ovoid tubercles up to 3.5 mm long. Nut 1.3–3.5 mm, trigonous, broadest about one-third from the base, brown. $2n = 60*$ (Dempsey *et al.* 1994). Flowering from late May to November.

Native. Found throughout British and Ireland (map, p.175) on cultivated land, in pastures, waste places and farmyards, by roads and tracks, and on coastal shingle beaches, sand-dunes and the mud of tidal rivers. It can endure very wet conditions, e.g. pond-sides or brackish conditions (as on the coast); it is, however, usually absent from shady places and acid moorland and heaths.

Widespread in Europe and Asia and introduced to North America and temperate regions worldwide as a weed. The species was monographed by Cavers & Harper (1964, 1967a, 1967b). Variation in British plants was studied by Akeroyd & Briggs (1983a, b).

R. crispus is unlikely to be confused with any other British or Irish dock except perhaps (in northern Britain) 53 *R. longifolius,* and 60 *R. patientia.* It does, however, show great variation in leaf crisping, height, branching, panicle density, and features of the valves and tubercles. The characteristic features of many variants remain distinct in cultivation and three subspecies can be recognised.

1 Nut 1.3–2.5 mm; tubercles of inner valves in fruit distinctly unequal, sometimes only 1 present **a. subsp. crispus**
1 Nut 2.5–3.5 mm; tubercles of inner valves in fruit often subequal 2
   2 Stems usually not more than 100 cm; infructescence dense
                                                       b. subsp. littoreus
   2 Stems usually 100–200 cm; infructescence lax **c. subsp. uliginosus**

Rumex crispus subsp. crispus 61a (A)
subsp. littoreus 61b (B)
f. unicallosus (C)

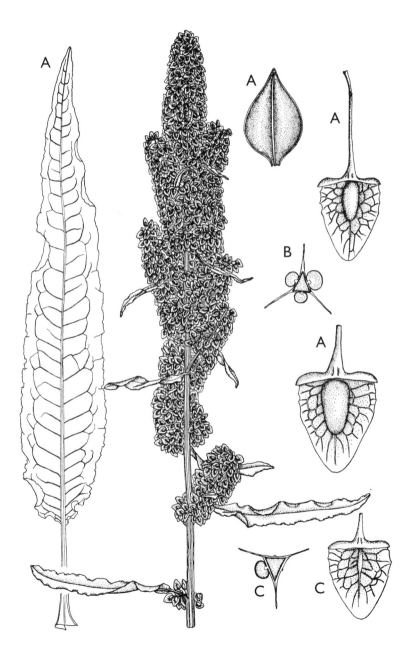

a. subsp. **crispus**

Annual, biennial or short-lived perennial, usually flowering in first year of growth. Leaves variably but often strongly crisped. Stems 40–150 cm. Panicle branched, lax or moderately dense in fruit. Tubercles 3, sometimes 1 (rarely 2), not always conspicuous, up to 2.5 mm, usually distinctly unequal. Nut 1.3–2.5 mm. Flowering from late May to September, but sometimes until as late as November.

This is the common weed of arable and ruderal habitats, found throughout Britain and Ireland. The plant is widespread worldwide.

b. subsp. **littoreus** (Hardy) Akeroyd

var. *littoreus* Hardy; incl. var. *trigranulatus* Syme

Perennial, flowering in second year of growth. Leaves fleshy, strongly crisped, rather glaucous. Stems usually not more than 100 cm. Panicle with short branches, sometimes subsimple, dense in fruit, often fastigiate. Tubercles 3, conspicuous, up to 3.5 mm, subequal. Nut 2.5–3.5 mm. Flowering in late June and July, usually slightly later than the other two subspecies.

Widespread and locally abundant around the coasts of Britain and Ireland, especially on shingle-beaches but also on sandy and rocky shores, natural and artificial earth banks by the sea, sand-dunes and the upper levels of saltmarshes. It is most characteristically a plant of strand-line communities on coastal shingle, together with species such as *Crambe maritima*, *Glaucium flavum*, *Lathyrus japonicus* subsp. *maritimus* and *Silene uniflora*. Where coastal habitats have been disturbed this subspecies forms mixed or intermediate populations with subsp. *crispus*. Plants from sand-dunes are also often intermediate between the two subspecies.

There are a few records of subsp. *littoreus* from the coasts of W. Europe, where it is probably under-recorded.

This subspecies undoubtedly forms a number of hybrids on disturbed ground by the sea. *R. crispus* subsp. *littoreus* × 73 *R. pulcher* has been recorded on stable coastal shingle at Aldeburgh, E. Suffolk (v.c. 25), and from W. Kent (v.c. 16).

c. subsp. **uliginosus** (Le Gall) Akeroyd

var. *uliginosus* Le Gall; var. *planifolius* auct. brit., non Schur; *R. elongatus* auct. brit., non Guss.

Perennial, with a massive taproot, sometimes not flowering until third year of growth. Leaves narrowly lanceolate, more or less uncrisped. Stems 100–250 cm. Panicle lax in fruit, with long branches. Tubercles 3, conspicuous, up to 3.5 mm, usually subequal. Nut 2.5–3.5 mm. Flowering in late May and June, slightly earlier than the other two subspecies.

A sporadic but locally very abundant constituent of plant communities on tidal mud and levees of rivers and creeks in southern England, Wales and southern and western Ireland. It usually occurs towards the upper limits of tidal influence, for example at the first bridge away from the sea, especially where the banks of the river are steep and unstable. Associated species include *Agrostis stolonifera* var. *palustris*, *Cochlearia anglica*, *Phragmites australis* and a robust, perennial variant of *Aster tripolium*. The largest populations of this variant are probably those in Co. Limerick (v.c H8) and Co. Clare (v.c. H9) on the tidal reaches of the River Fergus and smaller tributaries of the River Shannon to the west of Limerick City. It is also locally common on the larger rivers of S.E. Ireland, along stretches of the River Wye, notably at Tintern, Monmouth (v.c. 35), and smaller rivers flowing into the Bristol Channel, and on the River Medway in Kent (v.cc. 15–16).

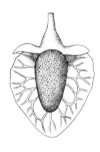

Probably endemic to N.W. Europe. Outside Britain and Ireland, subsp. *uliginosus* is known only from similar habitats in southern Brittany (Département de Morbihan), from where it was originally described as a variety.

Plants from the tidal banks of the River Thames in London (Surrey and Middlesex, v.cc. 17 and 21), collected as *Rumex elongatus* Guss. in the late 19th and early 20th centuries but now apparently extinct, are similar to subsp. *uliginosus* but are generally shorter and have very narrow, uncrisped leaves, a leafy inflorescence and valves often with only one tubercle. The plants seem to have disappeared after stretches of river-bank were rebuilt. In Ireland, similar plants occur in riparian communities at the furthest tidal limits of the River Barrow around St Mullin's, Co. Carlow (H13) and elsewhere. They merit at most varietal rank.

179

Hybrids

R. crispus forms hybrids with several species, especially with 55 *R. longifolius* in northern Britain (see under *R. longifolius*) and 74 *R. obtusifolius* (see 62 *R.* × *pratensis*).

R. × *schreberi* Hausskn. (61 *R. crispus* × 58 *R. hydrolapathum*): see under 58 *R. hydrolapathum*.

R. × sagorskii Hausskn. (61 *R. crispus* × 65 *R. sanguineus*)

Intermediate in habit between the parents. Leaves narrowly acuminate and usually narrowed to the base, with the margins conspicuously crisped. Panicle lax, rather leafy, with long branches; the broadly lingulate valves usually with a single large tubercle. Fertility low. Widespread on roadsides, hedge-banks and woodland margins in southern England from E. Cornwall (v.c. 2), S. Devon (v.c. 3) and S. Somerset (v.c. 5) to E. Sussex (v.c. 14) and W. Kent (v.c. 16); more local in Wales and N. Britain, extending north in scattered localities to Main Argyll (v.c. 98). Recorded in Ireland only in W. Mayo (v.c. H27) (Kitchener 1996a), and Co. Wexford (v.c. H12), but probably overlooked elsewhere. Known from a number of countries in W. Europe east to Sweden.

R. × *celticus* Akeroyd (61 *R. crispus* × 66 *R. rupestris*): see under 66 *R. rupestris*.

R. × pseudopulcher Hausskn. (61 *R. crispus* × 73 *R. pulcher*)

A variable plant, most frequently with the ascending branches and upright habit of *R. crispus*, but readily distinguished by the raised reticulations and 3 warty tubercles on the valves, which are as broad as in *R. pulcher*. Fertility low. Rarely the habit is bushy with the branches spreading at an angle of *c*. 90 from the stem, as in *R. pulcher*, but with much shorter teeth on the margins of the irregularly produced valves. In Britain known from a few dry, warm sites in rough or disturbed grassland from Cornwall (v.cc. 1–2), S. Devon (v.c. 3) and S. Somerset (v.c. 5) to W. Kent (v.c. 16) and E. Suffolk (v.c. 25), extending to Pembroke (v.c. 45). Reported in Ireland only from disturbed pasture on Sherkin Island, W. Cork (v.c. H3) (Akeroyd *et al.* 2011). Also known from Greece, Bulgaria and Spain.

R. × *pseudopulcher* on stable coastal shingle at Aldeburgh, E. Suffolk (v.c. 25), has *R. crispus* subsp. *littoreus* (Hardy) Akeroyd as a putative parent (see above).

R. × **areschougii** G. Beck (*R. heteranthos* Borbás) (61 *R. crispus* × 77 *R. palustris*)

Plant with characters of both parents in different combinations. The crisped leaf-margins of *R. crispus* are often evident, but the leaves are narrow and the panicle usually shows the arcuate-ascending branches and long, leafy bracts of *R. palustris*. The valves combine the broader, more triangular outline of *R. crispus* with the sometimes long teeth of *R. palustris*; one tubercle is larger. Fertility often low. Known from cut-over peat and pond margins in N. Somerset (v.c. 6) (Green, Green & Crouch 1997) and margins of lakes and reservoirs in Essex (v.cc. 18-19), Middlesex (v.c. 21) and Leics (v.c. 55), but undoubtedly under-recorded. Widespread through much of Europe.

R. × **fallacinus** Hausskn. (61 *R.crispus* × 78 *R. maritimus*)

Similar in general habit to *R. crispus* but with infructescence golden- rather than reddish-brown or brown. Basal leaves narrow, crisped. Panicle with numerous crowded, ascending branches. Valves intermediate between those of the parents, superficially similar to those of 62 *R.* × *pratensis*, but with longer teeth; pedicels slender. The plant also has the characteristically short anthers (< 1 mm long) of *R. maritimus*. Fertility low. In Britain recorded only from beside pools in grazing marshes between the Medway and Thames estuaries, W. Kent (v.c. 16) (Kitchener 1996b), on the partially broken-down bank of a drainage canal in Cambs (v.c. 29), and on a reservoir edge in Warks (v.c. 38). Also in C. Europe north to Sweden.

62 R. × pratensis Mert. & Koch

R. × *acutus* auct., non L. (61 *R. crispus* L. × 74 *R. obtusifolius* L.)

Plant robust, leafy, often regenerating well into the autumn. Basal leaves usually showing the crisped and undulate margins of *R. crispus*, but broader in proportion to their length and subcordate at the base, usually with the hairy midrib and petioles of *R. obtusifolius*. Stem leaves similarly resembling those of the two parents and more often closer to *R. obtusifolius*, especially in being papillose-hairy. Valves similarly intermediate in shape, each margin having an average of 4 teeth that are much shorter than those of *R. obtusifolius*; tubercles 1–3. Panicles, which produce a high proportion of infertile seed, regenerating from the base and lower nodes, and continuing to grow and flower over a long period, often turning orange or red. $2n = 44$–56 (Lousley & Williams 1975). Flowering from June to October.

The commonest of all dock hybrids in Britain and Europe, and the most fertile, *R*. × *pratensis* is widespread and it can be conspicuous wherever the parents occur together. This hybrid often forms a high proportion of a ruderal dock population. The plants exhibit the characters of the parents in widely varying proportions so that they range from plants near *R. crispus* to plants that are almost indistinguishable from *R. obtusifolius*. It occurs throughout Britain and most of Ireland (where it is probably under-recorded), in fields and waste places, by roadsides, on river-banks and on other disturbed ground. It is sometimes found in the absence of one or other parent.

Widespread in Europe; also recorded from the USA.

Field observations and measurements suggest that introgressive hybridisation between *R. crispus* and *R. obtusifolius* may occur extensively in Britain and Ireland (map, p.185) as well as elsewhere in Europe (Ziburski *et al.* 1986). Back-crossing occurs with both parents, but is probably commoner with *R. crispus* than with *R. obtusifolius* (Williams 1971). Back-crosses with *R. crispus* have been produced experimentally but have not yet been confirmed with the other parent. Hybrid plants are much less fertile than the parents. Experimentally the loss of fertility has been found to be about 20%, but it is undoubtedly much higher under natural conditions.

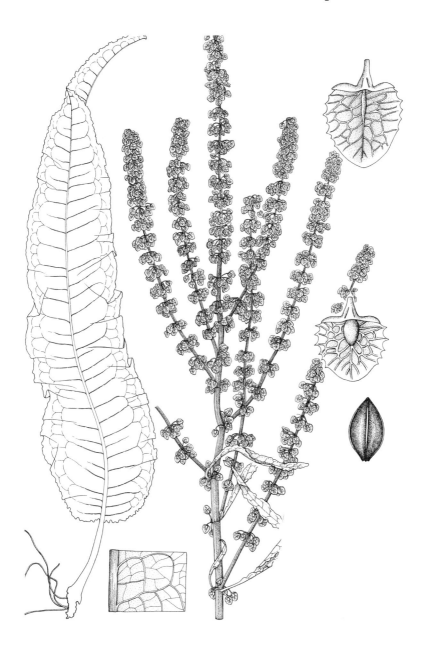

63 Rumex stenophyllus Ledeb.

A stiffly erect perennial 20–60(–120) cm tall, brownish or reddish, leafy, with the habit of *R. crispus*. Basal leaves oblong-lanceolate, often truncate at the base; apex rather obtuse; margins undulate but not as crisped as those of *R. crispus*. Stem leaves lanceolate, narrowing towards both ends. Panicle dense, with ascending branches, leafy below. Valves 3.5–4(–5) × 4–5 mm, orbicular, cordate at the base, reddish-brown, with a short triangular apex, the margins with 4–6 short, broad-based teeth 0.5–1 mm long; all valves bearing small oblong tubercules near the base. Nut *c*. 2 × 1.5 mm, apiculate at both ends, dark brown. $2n = 60$ (Pólya 1950). Flowering in July and August.

Native from N. and C. Asia to E. Europe, westwards to Austria and Slovenia; reported as an adventive in Britain, Scandinavia and Netherlands, probably introduced with grain, also in North America as widespread ruderal. In its native range it is mostly a plant of riparian or saline habitats. In Britain recorded with certainty only from Hordle, S. Hants (v.c. 11) and Avonmouth Docks, W. Gloucs (v.c. 34), where it persisted for some years. It has not been seen recently.

R. stenophyllus superficially resembles 61 *R. crispus* in the shape and undulation of the leaves, but the presence of rather neat regular serrations on the orbicular (not orbicular-deltate) fruiting valves should prevent confusion. The dense panicle somewhat suggests that of 55 *R. longifolius*, though it is less leafy and the fruiting valves bear tubercles.

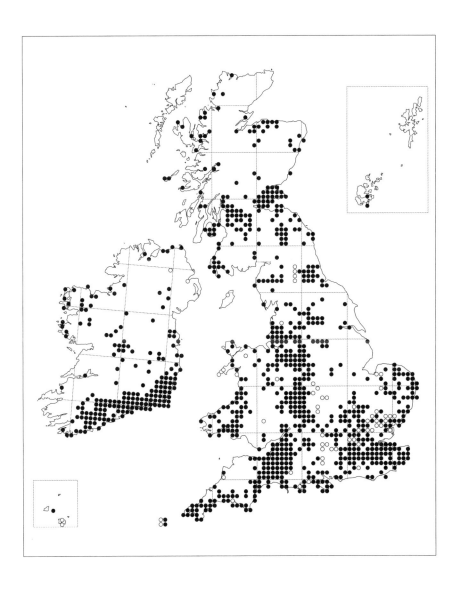

A perennial up to 30–60 cm tall, with several flexuous flowering stems from each rootstock. Basal leaves 10–30 × 2.5–6 cm, dull-green, of medium texture, oblong-lanceolate to ovate- or obovate-lanceolate, sometimes panduriform when young, abruptly rounded or subcordate at the base, subacute, glabrous, almost entire; petiole often equalling or exceeding the lamina. Stem leaves smaller, lanceolate, subacute, the margins sometimes crenulate. Panicle lax, leafy and open with numerous simple branches making an angle of 30–90° with the main stem, the flowers in rather remote whorls of 10–30, subtended by lanceolate or ovate bracts for about 2/3 of the length of the branches; flowers towards the end of the branches without bracts. Valves 2–3 × 1–2 mm, oblong-ovate or -lanceolate, with a long parallel-sided obtuse apex, entire, all 3 valves with an oblong, swollen tubercle 1–2 mm long, often more than half the length of the valve; fruiting pedicel about as long as or slightly longer than the perianth. Nut 1.5 × 1 mm, trigonous with acute angles, broadest near the base, acute, reddish-brown. $2n = 20$ (Jaretzky 1928). Flowering from July to October.

Native. In damp or marshy meadows, on the margins of ponds, ditches and streams, and beside muddy paths and field gates, particularly where water stands long in winter; also in brackish conditions. It sometimes forms a strand-line on the margins of ponds and lakes at the sides and levels to which fruits drift in late autumn. Throughout Britain and Ireland, but local or rare over much of Scotland (map, p.190).

Native in Europe east to the Crimea, N.W. Africa, Siberia and C. and S.W. Asia; naturalised in North America.

R. conglomeratus and 65 *R. sanguineus* grade into one another through a series of fertile intermediate variants, probably as a result of introgressive hybridisation. These intermediates occur commonly, especially in England, mainly where the ecological ranges of the two species overlap (Rackham 1961). It has been suggested that the two species might in fact constitute a single, very variable species. However, in S. Europe and elsewhere they remain quite distinct and do not intergrade; also their F1 hybrid (*R.* × *ruhmeri*: see below) shows reduced fertility.

A reliable character to separate the two species is the length of the fruiting pedicel – about as long as the valve or slightly longer in *R. conglomeratus* but much longer than the perianth in *R. sanguineus* (but see notes under 65 *R. sanguineus*).

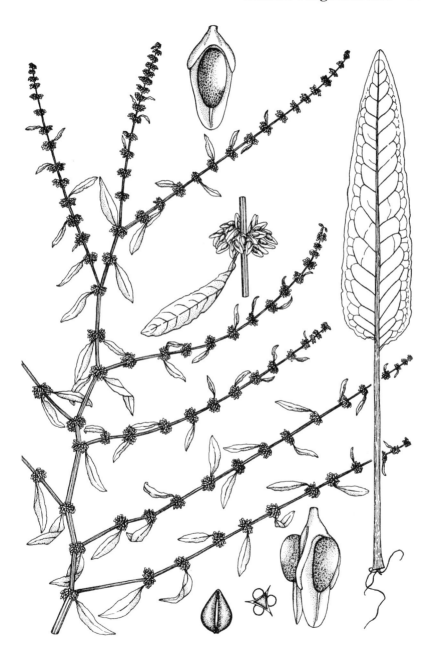

Hybrids

In addition to the hybrids described below, *R. conglomeratus* forms hybrids with 52 *R. cuneifolius* and 58 *R. hydrolapathum*: see under those species.

R. × **schulzei** Hausskn. (64 *R. conglomeratus* × 61 *R. crispus*)

Leaves narrowly lanceolate, slightly crisped. Panicle erect, with ascending branches (suggestive of 65 *R. sanguineus*) and leafy bracts. Valves intermediate in outline between those of the parents, i.e. more elongate than in *R. crispus* and with some slightly elongate tubercles. It is usually highly infertile. Hybrids involving *R. conglomeratus* rarely show evidence of widely spreading branches, which suggests that this character may be recessive.

Common and widely distributed about ponds and ditches, in disturbed grassland and on waste ground, mostly in England and Wales; in Ireland recorded only from West Cork (v.c. H3), Co. Waterford (v.c. H6) and Co. Kilkenny (v.c. H11), but perhaps overlooked. Widespread in Europe and recorded from N.W. Africa.

R. × **ruhmeri** Hausskn. (64 *R. conglomeratus* × 65 *R. sanguineus*)

Intermediate between the parents in leaf shape and the leafiness and density of the inflorescence and in having 1–3 tubercles on the valves. The tubercles are usually elongate rather than globose and the branches ascending rather than spreading. It can be rather scruffy and is variable in fertility but often semi-sterile. In Britain probably widespread and locally frequent. Authentic material has been seen from W. Cornwall (v.c. 1), S. Devon (v.c. 3), Somerset (v.cc. 5 and 6), S. Hants. (v.c. 11), W. Kent (v.c. 16), Oxon (v.c. 23), Worcs. (v.c. 37), Warks. (v.c. 38), Carms. (v.c. 44), and N.E. Yorks. (v.c. 62); in Scotland only from Main Argyll (v.c. 98) and Arran, Clyde Is. (v.c. 100); in Ireland from West Cork (v.c. H3), Co. Wicklow (v.c. H20), Meath (v.c. H22) and West Mayo (v.c. 27). Recorded from several European countries.

R. × *rosemurphyi* Holyoak (64 *R. conglomeratus* × 66 *R. rupestris*): see under 66 *R. rupestris*.

R. × **muretii** Hausskn. (64 *R. conglomeratus* × 73 *R. pulcher*)

Usually similar in habit to *R. pulcher* in the divaricate branching, but taller and rather leafier. Basal leaves often slightly panduriform. Valves with the strong reticulation of *R. pulcher* but smaller, lingulate, entire or with a few small teeth, and with tubercles less warty. Possibly frequent in disturbed rough grassland in southern England where the two parents grow together, though *R. pulcher* grows in dry sunny places and *R. conglomeratus* mostly grows in damper habitats.

There are records from Cornwall and Scilly (v.cc. 1–2), S. Devon (v.c. 3), N. Somerset (v.c. 6), Dorset (v.c. 9), Hants (v.cc. 11-12), Sussex (v.cc. 13–14), Kent (v.cc. 15–16), Surrey (v.c. 17), E. Suffolk (v.c. 25), Cambs (v.c. 29) and Monmouth (v.c. 35). In Ireland it has been found in disturbed pastures near the sea on Sherkin Island, West Cork (v.c. H3) (Akeroyd *et al.* 2011). Rare in Europe; also recorded from the Balkans, Eastern Aegean and Tunisia.

R. × **abortivus** Ruhmer (64 *R. conglomeratus* × 74 *R. obtusifolius*)

Similar to *R. obtusifolius* in habit. Lower leaves broad, obtuse or acute, subcordate at the base, papillose-scabrid on veins beneath. Panicle lax, with distant whorls and leafy bracts, especially below. Valves variable, even on the same plant, but usually lingulate, with a few conspicuous teeth and a single well-developed tubercle. Fertility is low. Widespread and often frequent in waste places and damp pastures, from southern England to central Scotland, but rarer in the north. Known in Ireland only from disturbed pasture on Sherkin Island, West Cork (v.c. H3) (Akeroyd *et al.* 2011) and waste ground in Knockboy, Co. Waterford (v.c. H6). Widespread in Europe.

R. × **wirtgenii** G. Beck (64 *R. conglomeratus* × 77 *R. palustris*)

Like a bushy *R. palustris* but with somewhat broader leaves. Valves with a lingulate apex, a few long, stiff teeth and 3 elongate tubercles. In Britain it is rather rare, by ditches, peat-cuttings, ponds, reservoirs and canals in southern and central England. Authentic material has been seen from N. Somerset (v.c. 6), W. Kent (v.c. 16), Surrey (v.c. 17), S. Essex (v.c. 18), Middlesex (v.c. 21), E. Norfolk (v.c. 27), Bedford (v.c. 30), Worcs. (v.c. 37), Leics. (v.c. 55) and Notts. (v.c. 56). Known from a few other countries in Europe.

R. × **knafii** Celak. (64 *R. conglomeratus* × 78 *R. maritimus*)

Like *R. maritimus* in habit but leaves somewhat broader. Panicle rather lax, with long branches and numerous leafy bracts, not turning golden at maturity. Valves usually with 1–2 very slender teeth on each margin and sometimes an elongate tubercle; pedicels filiform as in *R. maritimus*. Fertility low. On the margins of ditches, peat-cuttings, ponds and lakes, mostly in southern and eastern England but extending northwards to Cheshire (v.c. 58) and S. Yorkshire (v.c. 61 and 63). Recorded also from Germany and Austria.

64 Rumex conglomeratus

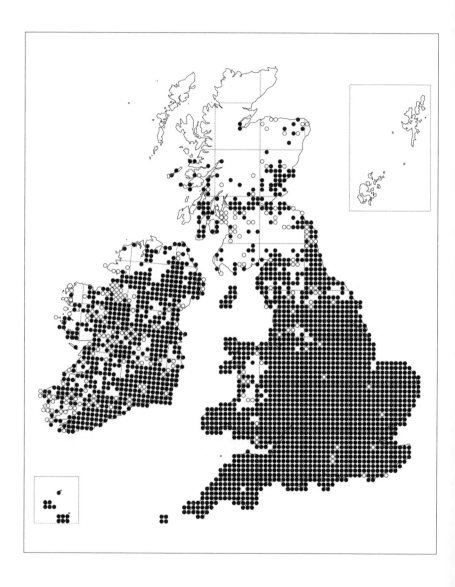

Rumex sanguineus (p.192) 65

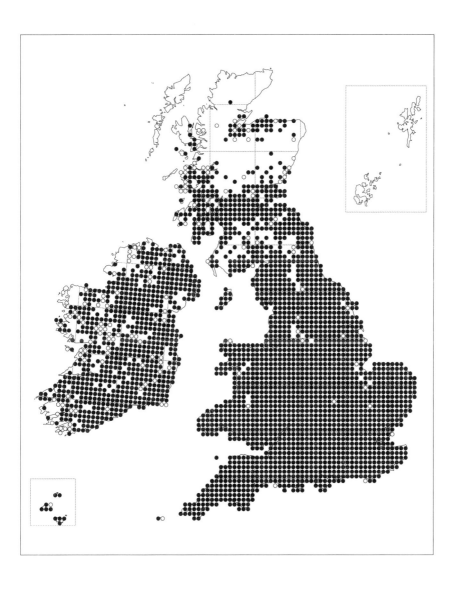

65   Rumex sanguineus L.

Wood Dock

A perennial 30–60 cm tall with several flowering stems. Basal leaves 15–30 cm, bright green, very thin in shady places, leathery in exposed situations, oblong (rarely subpanduriform), cordate or truncate at the base; apex acute or subacute; margins slightly undulate or crenulate; petioles 1/3–1/2 as long as the lamina. Panicle strict with ascending branches making an angle of less than 20° (–45°) with the main stem, usually with many branches; pedicels 1–5 mm, jointed at or very near the base; flowers in worls spaced at intervals of 1 cm or more, only the lower ones sometimes subtended by bracts.Valves subequal, 2.5–3 × 1–2 mm, narrowly oblong, subobtuse, entire; one bearing a light brown or reddish globose tubercle *c*. 1 mm in diameter occupying almost the whole width but much less than half the length of the valve, the others without or with small tubercles; fruiting pedicel much longer than the perianth. Nut 1.25–1.5 × *c*. 1 mm, trigonous with acute angles, broadest near the base, acute, dark brown. $2n$ = 20* (Montgomery 1997). Flowering in June and July.

R. sanguineus var. **viridis** (Sibth.) Koch (*R. condylodes* Bieb., *R. nemorosus* Schrader ex Willd.), the common native plant, has leaves with green veins and is a characteristic plant of woodland and hedgerow, thriving best in rides in woods on clay. It also occurs on roadsides and waste ground, as a coarser plant with thicker leaves. It is frequent throughout Great Britain and Ireland, but rare or absent over much of N. Scotland and the far west of Ireland (map, p.191).

Native in Europe north to Scotland and Scandinavia, N.W. Turkey and Caucasus; scarce in Mediterranean Europe.

R. sanguineus is frequently confused with 64 *R. conglomeratus*, and fertile intermediates are widespread, especially in England (see note under 64 *R. conglomeratus*). It is more slender, strict and erect than *R. conglomeratus*, with a much less leafy panicle and only one of the valves tuberculate, the tubercle being globose or subglobose and much shorter than the valves. It generally occurs in shadier and drier habitats.

R. sanguineus var. **sanguineus** (var. *purpureus* Stokes), Blood-veined Dock, is a striking plant with prominent blood-red or purple veins on the leaves, which long attracted the attention of gardeners and herbalists. It breeds true from seed and is known only in cultivation or as a rare escape. It sometimes persists, especially in the south, near houses, on roadsides and in copses and churchyards.

R. sanguineus hybridises with 61 *R. crispus*, 64 *R. conglomeratus*, 73 *R. pulcher* and 74 *R. obtusifolius*: see under those species. A spontaneous hybrid between *R. sanguineus* var. *sanguineus* and 73 *R. pulcher* was reported from the South London Botanical Institute garden (Lousley 1944).

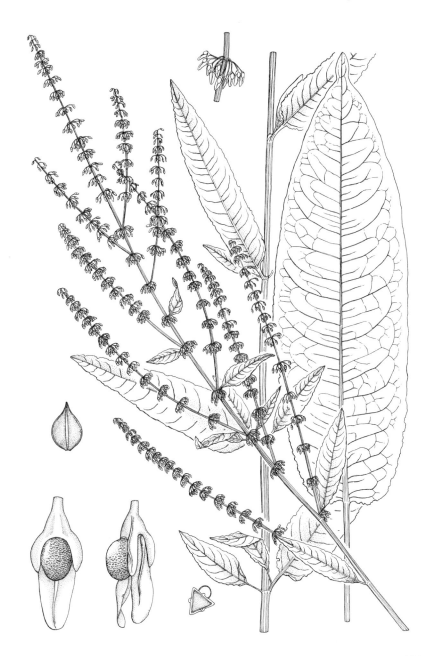

66 Rumex rupestris Le Gall   *Shore Dock*

A robust perennial 30–100 cm tall, with one to several stems arising from a woody stock. Basal leaves 10–30 cm, oblong- or broadly ovate-lanceolate, rather abruptly narrowed to a truncate or subcordate base, subobtuse, somewhat fleshy, dull green, glaucous, the margin entire, undulate; petiole usually less than one-third as long as the lamina. Stem leaves subsessile, oblong-lanceolate, gradually narrowed to the base, acute, the margin undulate, often crenulate. Panicle dense, the branches ascending, making an angle of about 45° with the main stem, with numerous short crowded branchlets and whorls of 5–20 flowers, usually crowded, often almost confluent, some below subtended by lanceolate bracts. Valves 3–4 × 2–2.5 mm, elongate-oblong, obtuse, not reticulate, entire, all bearing very large (*c.* 2.5 mm), very swollen, smooth elongate tubercles which occupy over two-thirds of the length of the valves and almost their entire width. Nut *c.* 2 × 1.5 mm, trigonous with acute angles, broadest near the rounded base, abruptly acute, reddish-brown. $2n = 20$ (Degraeve 1975a). Flowering in June and July.

Native in W Britain (map, p.197) on sandy and shingly seashores, often at the base of drift banks or sea-cliffs, where fresh water trickles onto the beach, and in damp dune-slacks. It is locally common in the Isles of Scilly (v.c. 1) and widespread but rare in Cornwall (v.cc. 1–2), with scattered stations in S. Devon (v.c. 3), Wales, in Glamorgan (v.c. 41), Pembroke (v.c. 45) and Anglesey (v.c. 52), and the Channel Islands (Guernsey, Herm and Jersey). It is extinct in N. Devon (v.c. 4) and almost so in Dorset (v.c. 9). Plants have been reintroduced to sites in Cornwall as part of a Species Recovery Programme carried out by English Nature and Plantlife.

Endemic to Atlantic coasts of W. Europe; its known distribution comprises N.W. Spain (Galicia), N.W. France, S.W. England and Wales. Threats to the species in Britain include the reconstruction and development of beaches, pressure from summer visitors, and an apparent increase in the severity of winter storms. Most populations number only a few individuals. *R. rupestris* is protected in Britain under Schedule 8 of the Wildlife and Countryside Act 1981, the EU Directive on the Conservation of Natural Habitats and of Wild Fauna and Flora 1992 (Habitats Directive), and the Bern Convention 1982. It is one of the world's rarest docks (IUCN global category: Vulnerable, UK category: Endangered). Daniels, McDonnell & Raybould (1998) have reviewed the distribution, status and genetic variation of the British populations.

This species should be looked for in south-eastern Ireland, although the specialised habitat may not be present.

Rumex rupestris 66

Hybrids

R. × **celticus** Akeroyd (66 *R. rupestris* × 61 *R. crispus*)

Variably intermediate between the parents. Leaves lanceolate, somewhat crisped. Best distinguished by the valves, which combine the broad outline of those of *R. crispus* with the more ligulate apex and very large, elongate tubercles of *R. rupestris;* margins entire or with 1–2 indistinct teeth. Some plants are so infertile that no valves develop, or there may be variation in tubercle size within an inflorescence. Plants can apparently also be fertile, with 3 exceptionally large, unequal, elongate ovoid tubercles similar to those of *R. rupestris*. These plants may resemble robust variants of *R. crispus*. This variation may indicate introgressive hybridisation. Most hybrids probably involve the coastal *R. crispus* subsp. *littoreus*.

Known from the Isles of Scilly (v.c. 1b), one site in W. Cornwall (v.c. 1), and one site in E. Cornwall (v.c. 2); formerly on sand-dunes at Kenfig, Glamorgan (v.c. 41). Unknown outside Britain.

R. × **rosemurphyi** Holyoak (66 *R. rupestris* × 64 *R. conglomeratus* Murrey)

Intermediate between the parents but taller and more robust. Panicles much-branched, with spaced whorls of flowers and numerous small bracts. Valves varying in size, oblong to oblong-lanceolate, blunt, with a single rounded tubercle. Fertility much reduced, although there is some evidence of introgressive hybridisation; $2n = 20*$ (Holyoak 2000). Known only from damp dune-slacks at Penhale, W. Cornwall (v.c. 1). Unknown outside Britain.

R. × **trimenii** E.G. Camus (66 *R. rupestris* × 73 *R. pulcher*)

More robust in habit than *R. pulcher*. Leaves broadly lanceolate, leathery. Panicle divaricately branched. Valves reticulately veined, with 3 warty tubercles. Known only on Samson, Isles of Scilly, and at Penhale, W. Cornwall (v.c. 1), and at Whitesand Bay, E. Cornwall (v.c. 2). Unknown outside Britain.

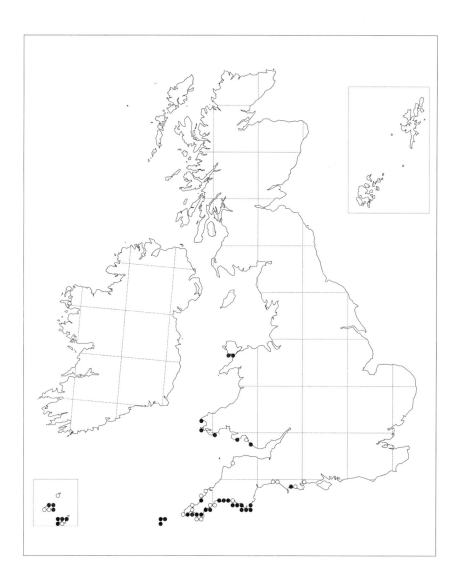

197

67 Rumex steudelii Hochst. ex A. Rich.

A moderately stout, glabrous perennial 40–120 cm tall. Basal leaves 10–20 ×
2–5 cm, oblong-lanceolate to lanceolate with a cordate base. Panicles divaricately branched, long and lax with bracts in the lower part only; whorls remote,
many-flowered; pedicels deflexed, longer than the flowering perianth, articulated
below the middle. Valves 3.5–5 mm long, ovate-cordate, the margins with 5-6
hooked teeth as long as the width of the valve; usually without tubercles, rarely
with a small tubercle on 1 or 3 valves. Nut 2.75–3 mm long, ovoid-triquetrous,
dark brown.

Native to the mountains of S.W. Arabia, Somalia and Ethiopia; also Southern
Africa. In Britain a rare casual introduced with wool-shoddy and recorded from
N. Hants (v.c. 12) and Beds (v.c. 30).

R. steudelii is closely related to 68 *R. bequaertii*, from which it differs in its
smaller upper leaves, more or less glabrous petioles and slightly larger valves
and nuts. The two may only be subspecifically distinct.

R. steudelii is also similar to 69 *R. nepalensis*, from which it can be separated
by its usually smaller size, narrower basal leaves and valves usually without
tubercles. The three species require further study across their whole range.

68 Rumex bequaertii J.J. De Wild.

An erect glabrous perennial up to 100(–150) cm tall. Leaves narrow, up to 30
cm long, but rarely exceeding 6 cm broad, oblong-lanceolate, often parallel-
sided, obtuse or sometimes rounded at apex, cuneate at base, glabrous or with
scattered papillae on the under surface, flat or crisped on the margins. Panicle
large and open, with long branches. Valves 3–4(–5) × 1.5–2.5 mm, elongate-
triangular, dark brown, with 3 (rarely 1) large tubercles, usually with an obtuse,
lanceolated apex, the margins with 5–6 strongly hooked teeth 1.5–2 mm long.
Nut 2.5–3 × 1.5–2 mm, ovoid, trigonous, glossy, brown. $2n = 40$ (Degraeve
1975a). Flowering from September to October.

Native of Africa from Ethiopia to Angola and South Africa (Transvaal); also in
Madagascar. In Britain a rare casual introduced with wool-shoddy, reported
from N. Hants (v.c. 12) and Kent (v.cc. 15–16).

Rumex steudelii 67 (A)
R. bequaertii 68 (B)
R. nepalensis (p.198) 69 (C)

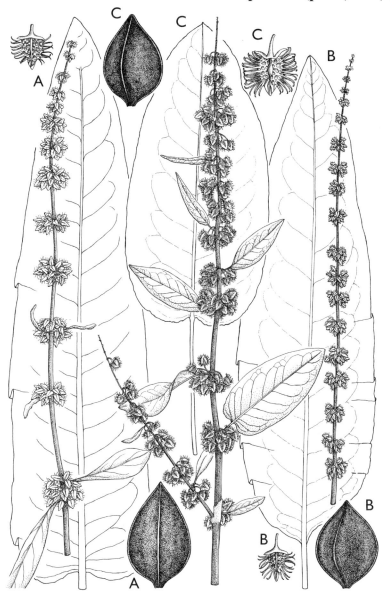

69 Rumex nepalensis Spreng. *(Illustrated on p. 197)*

A rhizomatous perennial 50–120 cm tall, branched above with spreading branches. Basal leaves large, oblong, ovate-oblong or triangular-ovate, cordate at the base, rounded or acuminate at the apex, pubescent beneath. Stem leaves subsessile, lanceolate, narrowed at the base. Panicle lax and open, leafy only at the base; whorls remote. Valves 4–7 × 3–5 mm, orbicular-ovate, one or all with an oblong tubercle, strongly reticulate, broadly winged, the wing pectinately toothed; the marginal teeth prominently hooked at the tip. Nut 4 mm long, dark brown. $2n = 120$ (Kihara & Ono 1926).

Principally a Himalayan species, but native from S.W. China to Anatolia and Palestine, extending west to a few localities in the mountains of the Balkan Peninsula and Italy. Its hooked valves adhere readily to the fleeces of sheep, and it was as a wool alien that it was recorded once in Britain at Tweedside, Selkirk (v.c. 79) in 1914 (Hayward & Druce 1919).

R. nepalensis is remarkable for the elegant hooked teeth that occur along almost the whole margin of the valves. In general habit it recalls 74 *R. obtusifolius* and should not be confused with occasional specimens of that species that have a few slightly hooked teeth on the valves.

70 Rumex brownii Campd. *Hooked Dock*

A perennial 25–60(–80) cm tall with a thick rhizome and several slender, rather flexuous, simple or slightly branched stems. Basal leaves 5–17 × 1–4 cm, narrowly ovate or lanceolate, usually constricted a little above the cuneate or cordate base, apex acute or acuminate, the margin finely crisped. Upper stem leaves much narrower and acute, often linear-lanceolate or hastate. Panicle lax, almost leafless, with 4- to 10-flowered, remote whorls. Valves 2.3–3.5 × *c.* 2 mm, deltate, reticulately veined, often reddish, not tuberculate, with 3–5(–6) teeth on each side usually more than half as long as the width of the valve, the midrib drawn out into an excurrent, hooked apex. Nut *c.* 2 × 1.25 mm, broadest at the middle, dark brown. $2n = 40$ (Johnson & Briggs 1962). Flowering from July to October.

Native of Australia, New Guinea and Indonesia; adventive in a number of European countries, mostly introduced with wool but perhaps also with cotton and bird-seed. In Britain it was first found on Tweedside in 1908 (Hayward & Druce 1919). Since then it has been found as a wool adventive in at least a dozen

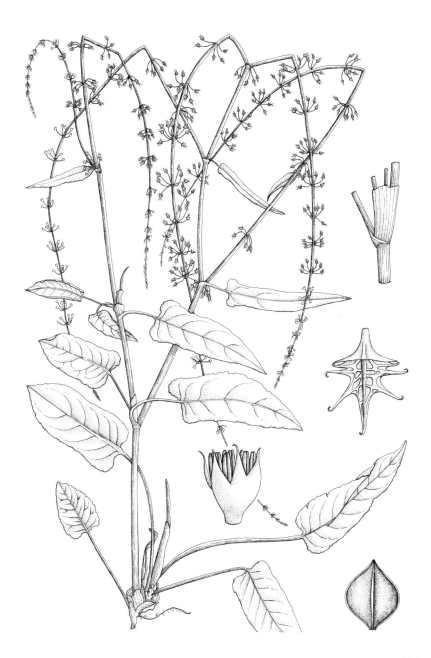

vice-counties from S. Devon (v.c. 3) and N. Somerset (v.c. 6) to E. Kent (v.c. 15) and northwards through Beds (v.c. 30) to S. Lancs (v.c. 59) and N.W. Yorks (v.c. 65); also persisting for many years on Tweedside, at Galashiels, Selkirk (v.c. 79), until at least 1966, and at Melrose, Roxburgh (v.c. 80), where it was recorded on river shingle in 1969.

71 Rumex crystallinus Lange

An erect, branched annual up to 40 cm tall. Leaves lanceolate or linear-lanceolate, cordate or with small auricles at the base, rarely tapering into the petiole, sometimes up to 15 cm long and very narrow on large specimens; ochreae membranous, conspicuous. Panicle with whorls many-flowered, distant or approximate, confluent towards the apex, and crisped leaves much longer than the whorls; pedicels 1.5 mm long. Valves small, 1.75–2.5 × 1 mm, lingulate, acuminate, with 1–2 long slender teeth on each side and a conspicuous tubercle on the midrib. Nut c. 1 × 5 mm, brown, broadest in the middle. $2n = 60$ (Ichikawa et al. 1971).

Native of Australia (New South Wales, South and Western Australia). In Britain a very rare wool adventive, reported reliably only from S.W. Yorks (v.c. 63) and N.W. Yorks (v.c. 65). Other records have been referred to 72 R. tenax.

72 Rumex tenax Rech. fil.

An annual of similar habit to 70 R. brownii but with narrower leaves and narrowly linear bracts. Panicle with up to 40 flowers, in remote whorls (internodes 3–4 cm). Valves 2.5–3 mm long, lingulate, acuminate, reticulate, with marginal teeth few, curved but not hooked; each valve with a prominent orange tubercle. Nut 2 × 1.25 mm, brown, broadest in the middle, gradually tapering to a fine point. $2n = 80$ (Ichikawa et al. 1971). Flowering in August and September.

Native of S. Australia. Introduced into Britain with wool, it was first found in 1911 by the River Tweed below Galashiels, Roxburgh (v.c. 80) (Hayward & Druce 1919). It was recorded in the 1960s from N. Hants (v.c. 12) in fields where wool-shoddy had been applied.

R. tenax is closely related to 70 R. brownii and 71 R. crystallinus; it superficially resembles 78 R. maritimus, but the very remote whorls and the fewer, shorter, curved teeth on the fruiting valves should prevent confusion.

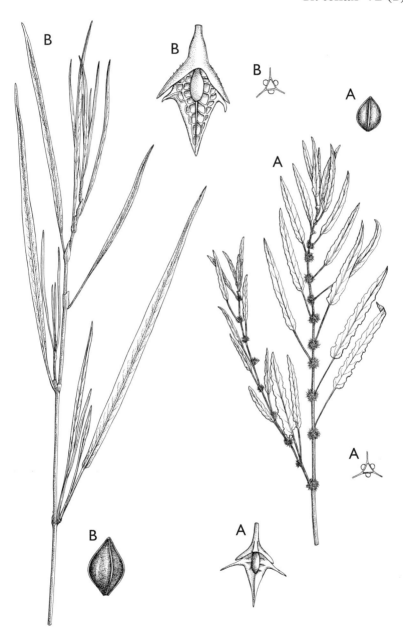

73a **Rumex pulcher** L. subsp. **pulcher** *Fiddle Dock*

A much-branched, somewhat untidy biennial or short-lived perennial, up to 50 cm tall but often shorter, with many crowns and several flowering shoots from the stock. Basal leaves 4–10 × 3.5–5 cm, panduriform or oblong, truncate or cordate at the base, obtuse, somewhat fleshy, dark green. Stem leaves few, oblong or lanceolate, subacute. Panicle with numerous simple or slightly branched, divaricate branches, making an angle of *c*. 90° with the main stem or, rarely, ascending; drooping in bud but becoming rigid in fruit, at maturity often forming a compact entangled mass. Panicle with whorls remote, a few with sub-sessile leafy bracts. Valves 6 × 3 mm, oblong- or ovate-triangular to ovate-orbicular, subcordate or truncate at the base, obtuse or subobtuse, reticulated with very prominent raised veins; margins on each side with 3–4 acute teeth which may be as long as half the width of the valve, all bearing oblong unequal tubercles which become warty when dry. Nut 2.5 × 1.5 mm, trigonous with acute angles, broadest near the truncate base, tapering to an acute apex, rather dark brown. $2n = 20$ (Shimamura 1929), 40 (Jaretzky 1928). Flowering from June to August.

Native. A plant of commons, village greens, churchyards, road-verges and dry pastures on sandy or chalky soils, most plentiful where the ground is somewhat disturbed (map, p. 207). It can persist in amenity grasslands and places where mowing regimes are relatively light. It is widespread and locally common in south-western and southern England, especially in dry grassland on or near the coast, becoming gradually less frequent north to Derbys (v.c. 57); in central England it is widespread but mostly on disturbed ground in places where it has perhaps been introduced.

In Ireland, *R. pulcher* was long regarded as a casual, but appears to be a rare native or long-established plant of dry grassland in southern and south-eastern coastal counties (Akeroyd 1993, Akeroyd *et al.* 2011). It is still present on Sherkin Island and Heir Island, West Cork (v.c. H3), where it has been known since 1896, and in at least two stations on the coast of Co. Wexford (v.c. H12).

Mediterranean region, S.W. Asia and S., W. and C. Europe; introduced in the USA and elsewhere.

The combination of divaricate, often tangled branches and valves with long teeth and rugose tubercles distinguishes this species from all other docks in Great Britain and Ireland. The panduriform leaves are distinctive, although not always present, especially in moister habitats, and occasional individuals of 64 *R. conglomeratus*, and 74 *R. obtusifolius* f. *pandurifolia*, exhibit this character.

Rumex pulcher subsp. pulcher 73a (A)
subsp. woodsii 73c (p.204) (B)

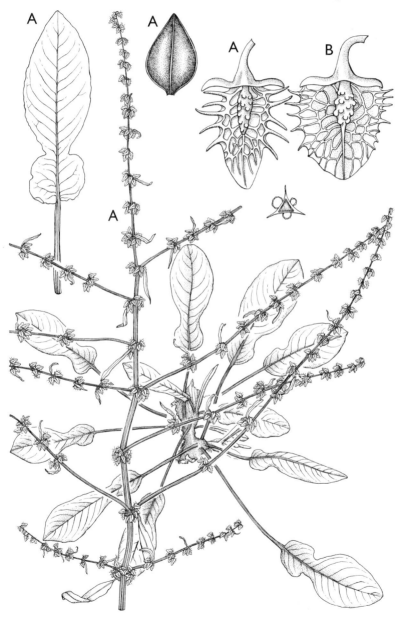

Two introduced subspecies (both with shorter teeth on the margins of the valves) have also been recorded, as rare casuals. Neither has been seen recently.

b. subsp. **anodontus** (Hausskn.) Rech. fil.

Basal leaves seldom panduriform. Margins of valves entire or with 1–2 teeth up to 0.5 mm long near the base. $2n = 20$ (Löve 1967).

Native of S.W. Asia and N.W. Africa. Introduced to Britain at Ware, Herts (v.c. 20), and, probably with grain, at Avonmouth Docks, W. Gloucs (v.c. 34).

c. subsp. **woodsii** (De Not.) Arcangeli   *(Illustrated on p. 205)*

subsp. *divaricatus* (L.) Murb.

Basal leaves sometimes panduriform. Valves broader than those of subsp. *pulcher*, suborbicular, the margins with up to 8 short teeth. $2n = 20$ (Löve 1986).

Native of the Mediterranean region and Turkey. Introduced to Britain more widely than subsp. *anodontus* at docks and on rubbish-tips, perhaps with bird-seed, also imported from the Southern Hemisphere as a wool adventive.

Hybrids

R. pulcher hybridises with 61 *R. crispus*, 64 *R. conglomeratus*, 66 *R. rupestris* and 74 *R. obtusifolius*: see under those species.

R. × mixtus Lambert (73 *R. pulcher* × 65 *R. sanguineus*)

Similar to *R. pulcher* in habit but leafier. Valves similar to those of *R. pulcher* but smaller and less strongly reticulate, with the margins entire or with a few short teeth and the tubercles unequal and less warty. Rare in dry grassland, with authentic specimens seen from Cornwall (v.cc. 1 and 2), S. Devon (v.c. 3), S. Somerset (v.c. 5), E. Sussex (v.c. 14), W. Kent (v.c. 16), Oxon (v.c. 23) and Cambs (v.c. 29). Also recorded from France and Romania.

Rumex pulcher 73

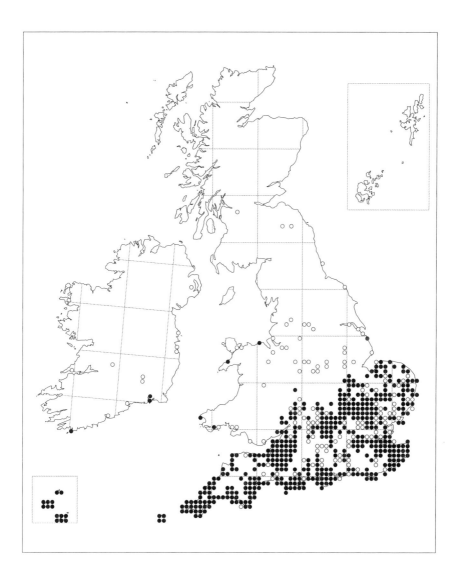

207

74 Rumex obtusifolius L.   *Broad-leaved Dock*

a. subsp. **obtusifolius**

A robust perennial 40–120(–150) cm tall, with a stout taproot and usually numerous crowns. Basal leaves 20–40 × 10–15 cm, ovate-oblong, strongly cordate at the base, obtuse or subobtuse, the margins entire or almost entire, glabrous above, usually papillose-scabrid on the veins beneath, dark green. Stem leaves smaller, narrower, more acute, often truncate or acuminate at the base. Panicle lax, with numerous single branches making an angle of about 45° with the main stem, leafy below. Valves 5–6 × 2–3 mm, ovate-triangular, obtuse, one (or sometimes all) with 1 prominent, elongate, smooth tubercle (rarely 3, f. *trigranis* (Danser) Rech. fil.); margins with 3–5 prominent teeth up to half the width of the valve (rarely longer, f. *subulatus* (Rech.) Rech. fil.); pedicels slender, up to 2.5 times as long as the valves. Nut *c.* 2 × 1.5 mm, trigonous with acute angles, broadest at about a third of the way from the base, acute at apex, narrowed rather abruptly to the base, medium brown. $2n = 40$ (Kihara & Ono 1926). Flowering from June to October.

Native. The commonest dock, often abundant on waste ground, field borders, hedge-banks, riverbanks, roadsides, pond- and lake-sides and in ditches and on cultivated ground. Throughout Britain and Ireland. Seedlings probably need open habitats to become established, but mature plants frequently persist in closed habitats under conditions of severe competition.

R. obtusifolius occurs throughout Europe and in Turkey and the Caucasus; subsp. *obtusifolius* has been introduced to North America and temperate regions worldwide as a weed. The species was monographed by Cavers & Harper (1964).

A very variable species; in E. and S.E Europe, subsp. *obtusifolus* is replaced by three other subspecies. Two of these are adventive in Britain.

1 One valve only tuberculate   **a.** subsp. **obtusifolius**
1 All valves tuberculate   2
   2 Leaves papillose-scabrid beneath; valves with several distinct teeth
                                           b. subsp. **transiens**
   2 Leaves glabrous beneath; valves entire or with a few short or indistinct
      teeth near the base   **c.** subsp. **sylvestris**

b. subsp. **transiens** (Simonk.) Rech. fil.

Petioles and leaf-veins papillose-scabrid beneath. Valves *c.* 5 × 2–3 mm, ovate-triangular, obtuse or rarely subacute, with a few short teeth near the base; all valves tuberculate, one tubercle usually larger than the other two. $2n = 40*$ (Al-Bermani *et al.* 1993).

A native of C. Europe, extending to NE France, Scandinavia and the Balkans. In Britain (map, p.213) naturalised by the River Thames in and around London, especially about Putney, Surrey (v.c. 17), and Chiswick, Middlesex (v.c. 21), and elsewhere in the Thames valley; also recorded from S.E. Yorks (v.c. 61) and a few sites in Scotland. Probably overlooked. Not recorded in Ireland.

This subspecies is intermediate between subspp. *obtusifolius* and *sylvestris*. It can be difficult to draw a sharp division across the continuous variation. In its native range it occurs in damper and less ruderal sites than subsp. *obtusifolius*.

c. subsp. **sylvestris** (Wallr.) Celak.

Whole plant glabrous or subglabrous. Valves 3–5 × 2-2.5 mm, narrowly triangular or lingulate, entire or with a few very small teeth near the base; all valves tuberculate, the tubercles subequal. $2n = 40$ (Löve & Löve 1961).

A native of E. and S.E. Europe, extending west to Norway and N. Italy. In Britain naturalised in the Lea Valley from Broxbourne, Herts (v.c. 20), to Walthamstow, S. Essex (v.c. 18), and found elsewhere mostly as a casual. Not recorded in Ireland.

In its native range it is a plant of woodland margins, in glades and rides, and along streams.

A fourth taxon, subsp. **subalpinus** (Schur) Celak., has leaves glabrous or minutely papillose-scabrid beneath, and valves narrowly triangular, subacute, with several teeth, one valve with a tubercle. $2n = 40$ (Löve & Löve 1961). Native to streamsides and woodland margins in the Carpathians and mountains of the Balkan peninsula, Turkey and the Caucasus, it has not so far been reported from Great Britain or Ireland.

Other variants of *R. obtusifolius* include f. *pandurifolia* (Borbás) Beck, which has panduriform leaves like those of 73 *R. pulcher* and might be confused with that species. Plants with stems and leaf veins suffused with deep purplish-red have been called f. *purpureus* (Poiret) Lousley.

Rumex obtusifolius subsp. transiens 74b (A)
subsp. sylvestris 74c (B)

Hybrids

In addition to those described below, *R. obtusifolius* subsp. *obtusifolius* forms frequent hybrids with 61 *R. crispus* (see 62 *R.* × *pratensis*), and also with 52 *cuneifolius*, 54 *R. aquaticus*, 55 *R. longifolius*, 58 *R. hydrolapathum*, 59 *R. cristatus*, 60 *R. patientia* and 64 *R. conglomeratus*: see under those species.

R. × dufftii Hausskn. (74 *R. obtusifolius* × 65 *R. sanguineus*)

Usually intermediate between the parents, with long, ascending branches at an angle of *c.* 45° with the main stem. Basal leaves like those of *R. obtusifolius* but more acute; cauline leaves often acute. Fruits in rather dense but remote clusters. Valves elongate and often lingulate, more or less acute, with a few short teeth near the base and at least one somewhat elongate tubercle. Fertility usually low.

Widespread in southern England and Wales, on woodland margins and rides, in disturbed and rough pastures, on river-banks or waste ground, northwards to W. Norfolk (v.c. 28), W. Gloucs. (v.c. 34), Worcs (v.c. 37) and Denbys (v.c. 50); also scattered further north to Stirlings (v.c. 86), Dunbarton (v.c. 99) and Clyde Is. (v.c. 100). In Ireland recorded from E. Cork (v.c. H5), Co, Waterford (v.c. H6), Co. Wexford (v.c. H12), W. Mayo (v.c. H27) and Co. Derry (v.c. H40). Known from several S., W. and C. European countries north to S. Sweden.

R. × ogulinensis Borbás (74 *R. obtusifolius* × 73 *R. pulcher*)

A variable hybrid that can resemble either parent. Leaves papillose-scabrid on the veins beneath. Panicle usually more open than in *R. obtusifolius*, but the branches ascending rather than divaricate. Valves reticulate, with 3 warty tubercles. Fertility is variable. Scattered in dry grassland and waste places in a few southern English counties from Cornwall (v.cc. 1 and 2) to Surrey (v.c. 17), and Cambs. (v.c. 29); also in Worcs (v.c. 37), and probably more widespread. Recorded mostly in S. Europe.

R. obtusifolius f. *subulatus* (Rech.) Rech. fil., a scarce variant with long slender teeth on the margins of the valves, can be confused with this hybrid.

R. × steinii A. Becker (74 *R. obtusifolius* × 77 *R. palustris*)

A rather tall, much-branched, leafy plant, resembling *R. palustris* in the candelabra-like, ascending branches, the thin leaves, and long teeth on the valves. It resembles *R. obtusifolius* in the more robust habit, broader, more obtuse leaves, often with a cordate base and papillose-scabrid on the veins beneath, and the broader valves with more ovoid tubercles. Rare, recorded only from peat cuttings and pond margins in N. Somerset (v.c. 6) (Green, Green & Crouch 1997), a river bank in Cambs (v.c. 29), and a lake shore in Leics (v.c. 55), with pre-1960 records from Surrey (v.c. 17) and Middlesex (v.c. 21). N. and E.C. Europe.

R. × **callianthemus** Danser (74 *R. obtusifolius* × 78 *R. maritimus*)

A luxuriant plant with ascending branches and lanceolate leaves, often resembling *R. maritimus* but larger, less spreading and branched in habit, and brown rather than golden-yellow at maturity. Leaves similar to those of *R. obtusifolius* but more acute. Pedicels very slender. Valves much broader, like those of *R. obtusifolius* in outline but with unequal tubercles and numerous slender teeth along the margins. Fertility is variable.

Recorded from peat cuttings in N. Somerset (v.c. 6) (Green, Green & Crouch 1997), the sides of ditches, pits and ponds in Herts (v.c. 20) and Cambs (v.c. 29), disturbed ground near a stream in Berks (v.c. 22), lake shores in Leics (v.c 55), and Cheshire (v.c. 58), and a roadside and sewage farm in Worcs (v.c. 37). There is an 1892 specimen from Surrey (v.c. 17). A rare hybrid in Europe, known from Sweden to Austria, and perhaps more widespread in Britain.

Rumex obtusifolius subsp. transiens   74b

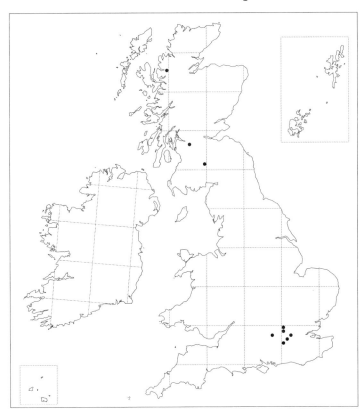

75 Rumex dentatus L. subsp. halacsyi (Rech.) Rech.fil.

Aegean Dock

An annual or sometimes biennial 20–70 cm tall, the stem erect, rather stiff, unbranched or with a few usually straight, ascending branches. Basal leaves small, 2–3 times as long as wide, lanceolate or (rarely) panduriform, truncate or subcordate at the base, weakly hairy or glabrous. Panicle long, clustered; whorls remote with linear-lanceolate bracts which have cuneate bases; pedicels longer than the valves (sometimes almost twice as long), jointed well below the middle, rather slender. Valves 4–6 × 2–3 mm, triangular, subacute, all or only one with a tubercle, yellowish-green ripening to brown; margins with 1–5 straight teeth 3–6 mm long. Nut 2–2.75 mm long. $2n = 40$ (Jaretzky 1928).

Native of Asia, extending to the Balkans and Libya; a weed or ruderal of river-banks, damp and cultivated ground. In Britain it has occurred as a rare and sporadic introduction with grain or wool-shoddy. In Ireland there are old records from Belfast (v.c. 38) (Reynolds 2002).

An extremely variable species. The taxonomy of *R. dentatus* requires further study, but Rechinger (1932, 1949) recognised seven subspecies, four of which have been recorded in Britain. As all but one (subsp. *nipponicus* (Franch. & Sav.) Rech fil., from E. Asia) of the seven have been recorded at some time in Europe, at least as casuals, an adaptation of his keys is given below.

1 Teeth of the valves distinctly longer than the width of the valves; valves *c*. 5 mm wide   2

  2 All valves bearing tubercles; nerves of the valves stout, with 2 reticulation masses on each side of the tubercle   subsp. **mesopotamicus**

  2 One tubercle developed, the other two rudimentary or absent; nerves of the valves slender, with 3 reticulation-masses on each side of the tubercle   subsp. **reticulatus**

1 Teeth of the valves shorter to as long as the width of the valves; valves less than 5 mm wide   3

  3 Valves narrow, usually entire or with 1–2 rather short teeth on each side; tubercles occupying almost the whole width of the valves, without any reticulation visible on either side   4

    4 Fruiting spike leafy; usually no reticulation (at most one mass) visible near the globose tubercle   subsp. **dentatus**

    4 Fruiting spike scarcely leafy; usually one reticulation-mass visible near the ovoid tubercle; plant turning black on drying   subsp. **nigricans**

Rumex dentatus

subsp. dentatus 75a (A), subsp. halacsyi 75b (B)
subsp. klotzschianus 75c (C), subsp. mesopotamicus 75d (D)
subsp. nigricans (see key) (E), subsp. reticulatus (see key) (F)

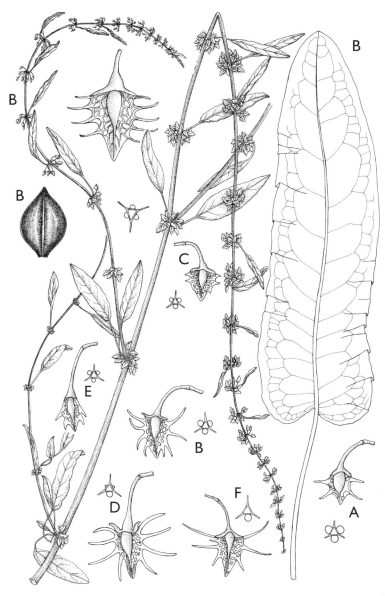

3   Valves relatively broad, usually with 3–4 teeth on each side; up to 3
    reticulations visible on either side of the tubercle   5
    5   Valves 3–4 mm wide, triangular in outline, teeth up to 3 mm long,
        strong; nerves prominent   subsp. **halacsyi**
    5   Valves *c*. 2.5 mm wide, more rounded in outline; teeth up to 2 mm
        long, more slender; nerves more delicate   subsp. **klotzschianus**

a. subsp. **dentatus**

subsp. *callosissimus* (Meisn.) Rech. fil.

$2n = 40$ (Löve & Löve 1961). Native from Syria and Palestine to Egypt and
Tunisia; known in Europe only as a casual. In Britain reported only from S.
Lancs (v.c. 59).

b. subsp. **halacsyi** (Rech.) Rech. fil.

$2n = 40$ (Jaretzky 1928). Native from Afghanistan to Turkey and S.E. Europe;
naturalised in S.W. Spain and perhaps elsewhere. The most frequently recorded
subspecies in Britain, including records from N.W. Yorks (v.c. 63) in the 1990s.

c. subsp. **klotzschianus** (Meisn.) Rech. fil.

$2n = 40$ (Löve 1967). Native of E. and S.E. Asia; known in Europe only as a
casual. In Britain reported from the Bristol area, N. Somerset (v.c. 6), and W.
Gloucester (v.c. 34).

d. subsp. **mesopotamicus** Rech. fil.

$2n = 40$ (Jaretzky 1928). Native from Iraq to Palestine, Arabia and Egypt;
known in Europe only as a casual. In Britain reported only from S. Hants (v.c.
11), S. Lancs (v.c. 59) and N.W. Yorks (v.c. 65).

Neither of the two other subspecies, subsp. *nigricans* (Hook. fil.) Rech fil.,
from N. India and Sikkim, and subsp. *reticulatus* (Besser) Rech. fil., from C.
Asia to Iraq and Ukraine, has been reported from Britain.

Hybrids

The hybrid with 78 *R. maritimus* was raised experimentally by Danser (1922d) but is unknown in the wild, apart from a possible record from Blackheath, south London (v.c.16), in 1920 (Lousley & Williams 1975). The hybrid with 61 *R. crispus* is known from Asia (Rechinger 1949) and the hybrid with 77 *R. palustris* from Hungary (Rechinger 1932).

76 Rumex obovatus Danser *(Illustrated on page 217)*

An erect annual or biennial 20–40(–70) cm tall, with a rigid stem. Upper branches making an angle of *c.* 45° with the main stem, the lower branches spreading, all leafy to the apex. Basal leaves up to 12 cm long and, like the lower stem leaves, obovate, obtuse, with a rounded or subcuneate base, papillose-scabrid on the veins beneath. Middle stem leaves narrower, cuneate at the base, subacute at the apex, with shorter petioles or sessile. Panicle with whorls remote below, somewhat congested at the extremities of the branches, all subtended by leafy bracts. Valves 4–5 × *c.* 3 mm, ovate, with subacute apex and 4–5 subulate teeth near the base on each margin, reticulate-rugose, all with large (*c.* 2.5 mm) ovate, subobtuse, warty tubercles. Nut *c.* 2.8 × 1.6 mm, pale or dark brown. $2n = 40$ (Löve 1967).

Native of Argentina and Paraguay in muddy and salty, periodically inundated places; reported as a former frequent casual in Fennoscandia (but rare after 1960) and other European countries, imported with grain. In Britain a rare introduction with wool-shoddy and grain on rubbish-tips, near grain mills and warehouses and about docks. Recorded from scattered localities from N. Somerset (v.c. 6) to Middlesex (v.c. 21) and S. Essex (v.c. 18), and northwards to Lanarks (v.c. 77) and Edinburgh, Mid-Lothian (v.c. 83). Much of the material seen is immature, and the species has not been recorded since the early 1970s.

R. obovatus shows little variation but can be mistaken for 73 *R. pulcher*, especially as the leaves are occasionally panduriform. It can be distinguished by the base of the cauline and radical leaves not being cordate, by the thicker and more leathery leaves, and by the much less spreading branches and the broader valves in less remote, more numerous-flowered whorls. It may be distinguished from 75 *R. dentatus* by its warty tubercles.

Hybrids

The hybrid with 61 *R. crispus* is the only one recorded from Britain.

R. × bontei Danser (76 *R. obovatus* × 61 *R. crispus*)

Superficially resembling *R. obovatus* but with the basal leaves oblong-ovate to obovate-lanceolate, with the margins undulate and crisped. Branches leafy below, with whorls many-flowered. Valves variably and irregularly toothed, with many fruits infertile (Danser 1926). A casual at Avonmouth Docks, W. Gloucs (v.c. 34), in 1928 (Sandwith & Sandwith 1936). Also recorded from Sweden.

Two other closely related species may have occured in Britain as casuals. Some immature British specimens of *R. obovatus* have been tentatively identified as **R. paraguayensis** D. Parodi, a native of Paraguay and Argentina, formerly adventive in Holland and Sweden (Rechinger 1933a). The species differs from *R. obovatus* by the smaller valves *c.* 3 × 2 mm and smooth tubercles. However, until adequate material is available, the species must be excluded from the British list. **R. violascens** Rech. fil., a native of the south-western USA, similar to *R. paraguayensis* but without regular axillary branches and with smaller valves 2.5–4 × 2–3 mm, was once reported as an adventive in Denmark (Rechinger 1937).

77 **Rumex palustris** Sm. *Marsh Dock*

A rather robust annual, biennial or short-lived perennial 10–60(–100) cm tall, with one or several shoots from the stock, the whole plant turning pale brown or reddish-brown as the fruits ripen. Basal leaves 20–35 cm long, rather thin, lanceolate or oblong-lanceolate, narrowed or subcordate at the base, narrowed gradually to a subacute apex, plane; margins entire; petioles usually about a quarter to a third the length of the lamina. Stem leaves narrowly lanceolate, very gradually narrowed at both ends, shortly stalked or subsessile, often crenulate near the base. Panicle candelabra-like, with numerous arcuate-ascending branches making an angle of 30–90° with the main stem, or plant almost unbranched in starved examples; whorls globose, with numerous flowers, distant below, congested above, nearly all with lanceolate leafy bracts. Valves 3–4 × 1.5–2 mm, narrowly lingulate, truncate at the base, acute, each with a subequal, oblong, light-brown or reddish, swollen tubercle 1.5 × 1 mm; margins with 2-3 stiff, bristle-like teeth, as long as the width of the valve; pedicels slender, as long as or up to one and a half times as long as the valves. Nut *c.* 1.5–2 × 1 mm, light brown, trigonous with acute angles, broadest above the middle, tapering to an acute apex. $2n = 40$ (Jaretzky 1928), 60* (Montgomery *et al.* 1997). Flowering from June to August.

Native. On wet, nutrient-rich mud beside ponds and dykes and in clay- and gravel-pits, marshes, peat-cuttings and other places inundated during the winter, in southern England and the Midlands, rarely further north (map, p.222). It a characteristic plant of recently abandoned peat-cuttings on the Somerset Levels, N. Somerset (v.c. 6) (Green, Green & Crouch 1997). In Ireland it is a very rare casual.

Native of W. and C. Europe, extending to C. Italy and C. Greece; rare in N.W. Africa and S.W. Asia.

This species is frequently confused with 78 *R. maritimus*, from which it can be distinguished in fruit by the pale brown or reddish-brown infructescence, the slightly thicker fruiting pedicels and valves with more obtuse tubercles and slender but somewhat rigid spines. In flower, especially visible in fresh material, the longer anthers (0.9–1.3 mm, compared with 0.4–0.6 mm in *R. maritimus*) are diagnostic.

Hybrids occur with 59 *R. cristatus*, 61 *R. crispus*, 64 *R. conglomeratus*, 74 *R. obtusifolius* and 78 *R. maritimus*: see under these species.

77 Rumex palustris

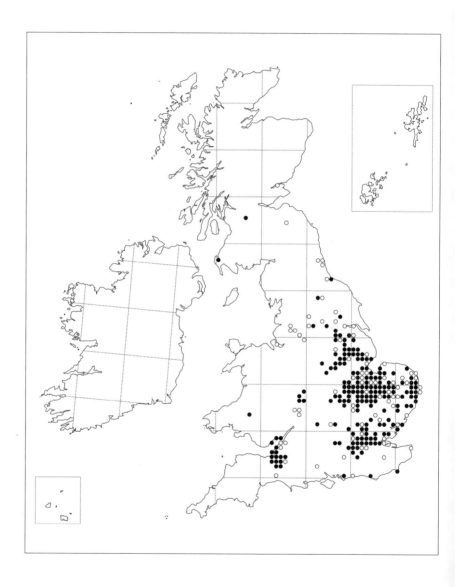

78 Rumex maritimus L. *Golden Dock*

An elegant, glabrous annual, biennial or short-lived perennial 20–50(–100) cm tall, the whole plant turning golden-yellow as the fruits ripen. Basal leaves 8–20(–40) cm, thin, linear- to oblong-lanceolate, usually very gradually narrowed at both ends, rarely subtruncate at the base, the margins entire, usually rather undulate; petiole about one-fifth the length of the lamina. Stem leaves narrowly lanceolate, very gradually narrowed at both ends, stalked or subsessile, the margins entire. Panicle dense, with numerous, often branched, ascending, rather leafy, branches making an angle of 30–60° with the main stem, rarely unbranched. Valves 2.5–3.5 × 1–2 mm, ovate to rhombic-triangular, cuneate at the base, acute, each with an elongate, narrow, pale yellow tubercle *c.* 1.5 × 0.75 mm; margins of valves with 2–5 very slender, flexuous teeth, 2–3 times as long as the width of the valve; pedicels filiform, somewhat longer than the valves. Nut tiny, 1–1.5 × *c.* 0.75 mm, trigonous with acute angles, broadest above the middle, subacute at the base, tapering to an acute apex, light brown. $2n = 40*$ (Montgomery *et al.* 1997). Flowering from July to September.

Native. On the margins of pools, lakes, rivers and dykes, in peat-cuttings, clay-pits and wet hollows in marshy fields, always in places inundated through the winter until late spring. In spite of its name, *R. maritimus* shows no preference for maritime conditions (map, p.223). Rare and generally decreasing, from southern England north to Salop (v.c. 40) and Cheviot (v.c. 68), with scattered records, mostly pre-1930, in southern and central Scotland north to Kintyre (v.c. 101) and Angus (v.c. 90).

In Ireland it is rare and sporadic, recorded recently only at a few sites: e.g. Fenit, N. Kerry (v.c. H2), Kilcolman Fen, E. Cork (v.c. H5), Lough Gur, Co. Limerick (v.c. H8), Lady's Island Lake, Co. Wexford (v.c. H12), Donabate, Co. Dublin (v.c. H21), and Loch Gara, Co. Sligo (v.c. H 28). It is protected in the Republic of Ireland by the Flora Protection Order 1987 (IUCN category, Rare).

Native of N. and W. Europe south to C. France and the Danube basin, N. Asia and, if 79 *R. fueginus* is included (see below), the temperate Americas.

This species is frequently confused with 77 *R. palustris*, from which it can be distinguished by the yellowish infructescence, the filiform fruiting pedicels and the valves with more acute tubercles and flexible, bristle-like spines on the margins, which give the infructescence a whiskery appearance. In flower, especially in fresh material, the shorter anthers (0.4–0.7 mm long, compared with 0.9–1.5 mm in *R. palustris*) are diagnostic. *R. maritimus* often stains herbarium sheets red.

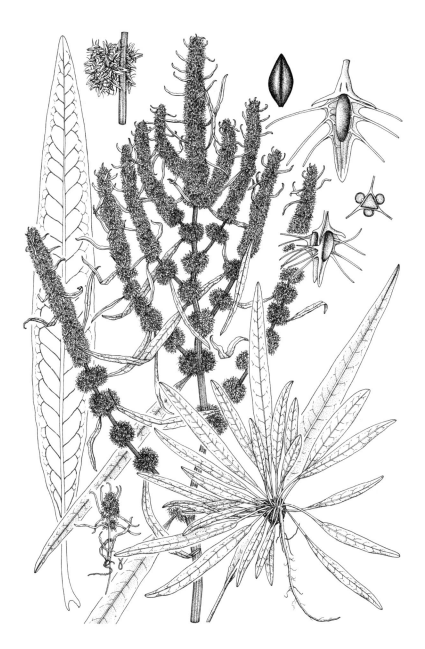

Hybrids

Hybrids occur with 59 *R. cristatus*, 61 *R. crispus*, 64 *R. conglomeratus* and 74 *R. obtusifolius*: see under these species.

R. × **henrardii** Danser (78 *Rumex maritimus* L. × 77 *R. palustris*)

Intermediate between the parents but often more robust; closer to *R. palustris* in general appearance and without the yellow colour of *R. maritimus*. Branches of panicle numerous, ascending, slender, leafy; pedicels more or less filiform. Valves poorly developed, with small tubercles and rather stiff bristle-like spines. Anthers 1–1.3 mm. Fertility very low, indeed plants mostly apparently sterile, correlating with the chromosome number difference reported between *R. maritimus* and *R. palustris* by Montgomery *et al.* (1997).

A rare hybrid, recorded in Britain in the Somerset Levels, in former marl pits on the Isle of Axholme, S. Somerset (v.c. 5), in old peat-cuttings and marl-pits, N. Somerset (v.c. 6) (Green, Green & Crouch 1997), and on wet mud of sewage farm settlement beds at Kidderminster, Worcs (v.c. 37); there is an unconfirmed 1930 record from Boarstall, Bucks (v.c. 24) (Druce 1930). Also recorded from the Netherlands, Denmark and Sweden.

J.E. Lousley (Lousley & Williams 1975) did not accept Druce's report of this hybrid from v.c.24, although based on a specimen sent to B.H. Danser (Druce 1930), who had described the hybrid (Danser 1916). A voucher specimen for the v.c. 5 record in Lousley's herbarium (Miss J. Gibbons [E. Joan Gibbons], 1949, **RNG**) is similar to plants collected in 1994 in v.c.6 by I. and P. Green.

79 Rumex fueginus Phil.

A reddish-brown plant similar to 78 *R. maritimus*. Extremely variable, it differs from *R. maritimus* in having the young stems covered with minute rough papillae, narrowly linear leaves that are truncate or subcordate at the base, and small valves 1.7–2 × 0.7–0.9 mm. $2n = 40$ (Mulligan 1959).

Native to North America and temperate South America; an occasional casual in N. Europe, especially about ports and mills. In Britain a very rare casual introduced with wool, confirmed from the Glasgow area and Tweedside by Rechinger (1949).

Closely related to 78 *R. maritimus*, with which it has been much confused, particularly in the USA. Indeed, some authors treat it at subspecific rank (*R. maritimus* subsp. *fueginus* (Philippi) Hultén).

80 Rumex bucephalophorus L. *Horned Dock*

A usually reddish annual with slender ascending stems, often spreading at the base. Basal leaves spathulate. Stem leaves small, ovate or lanceolate, cuneate at the base. Inflorescence a simple raceme; pedicels reflexed, much thickened in fruit towards the base of the perianth. Valves $2–4 \times 1–1.5$ mm, with 2–3 teeth on each margin; each valve with a minute swelling that is scarcely a tubercle. $2n = 16$ (Kihara & Ono 1926, Degraeve 1975b).

Native of the Mediterranean region, Iberian peninsula, Canary Islands, Azores and Madeira, where it is widespread and variable. In Britain a very rare casual, with a few old records from southern England, South Wales and southern Scotland, especially Leith Docks, Mid-Lothian (v.c. 83).

One of relatively few annual docks. Rechinger (1939) and Press (1988) have described the considerable variation in this species in its native range.

Dubious records of *Rumex* species

The following four docks have been recorded as adventive in Britain, but the records must be treated as dubious.

R. flexuosus Sol. ex Hook. fil.

Recorded from Selkirk (v.c. 79) (Hayward & Druce 1919). The specimen on which the record is based is immature and may be referable to *R. flexuosiformis* Rech. fil. (Lousley 1944), subsequently subsumed within *R. drummondii* Meisn. (Rechinger 1984). Recorded also from Mid-W. Yorks (v.c. 64) (Lees *et al.* 1942), but it is thought that the specimen was 70 *R. brownii* (Lousley 1944). Native to New Zealand; sometimes grown in gardens (Akeroyd 1989), so it may turn up again.

R. drummondii (Meisn.) Rech. fil. (*R. flexuosiformis* Rech fil.)

See note above under *R. flexuosus*. Native to Western Australia.

R. magellanicus Campd.

Closely related to 52 *R. cuneifolius*, from which it may be distinguished by its narrower and much more crisped leaves. The name was misapplied to some earlier gatherings of *R. cuneifolius* from Glasgow, Leith and Cardiff, but later, after specimens had been seen by B.H. Danser, deleted from the British list (Druce 1924, Hall 1937). The species appears to be rare even in its native Argentina and Chile and is an unlikely British adventive.

R. giganteus Aiton

Recorded from Oxford in 1918 in error for 57 *R. confertus* Willd. *R. giganteus* is a remarkable dock with a climbing habit and stems up to 12 m long in its native habitat, the forests of Hawaii. It is most unlikely ever to have occurred in Britain!

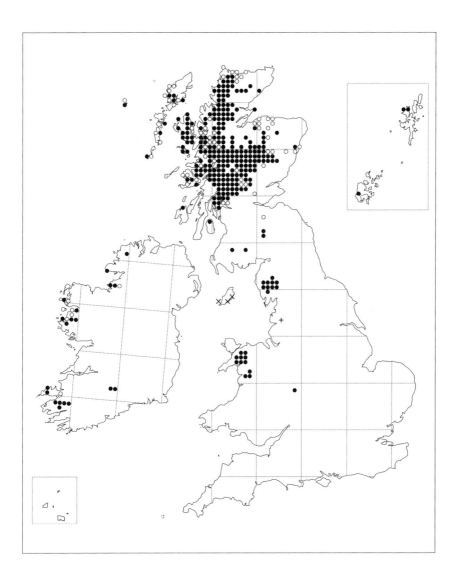

229

81 Oxyria digyna (L.) Hill

Mountain Sorrel

A tufted, glabrous perennial up to 40 cm tall but often much smaller, with a somewhat woody rootstock covered with persistent leaf-bases. Leaves 2–4 × 1–3(–6) cm, reniform, mostly basal, rather dull green, becoming reddish in late summer, acid to the taste. Inflorescence branched, leafless, mostly rather congested. Flowers dioecious, wind-pollinated. Tepals 4, without tubercles; the 2 outer spreading or reflexed, not accrescent, the 2 inner appressed to the nut, enlarging somewhat in fruit. Stamens 6. Stigmas 2. Nut 3–4 mm long, much longer than the inner perianth segments, lenticular, broadly winged, cordate at the base, with a small notch at the apex, green, often with a bright red margin, becoming scarious. $2n = 14$ (Kihara 1927). Flowering from July to September.

Native. On damp rock-ledges and by mountain streams, up to 1300 m in Breadalbane, Scotland, but sometimes carried downstream to sea-level on river shingle. It tolerates a range of soil pH. Widespread in the Scottish Highlands extending south to North Wales, the Lake District and Galloway; in Ireland, on mountains of the west, from Kerry (v.cc. H1–2) to Donegal (v.cc. H34–35) and in the Galty Mountains of S. Tipperary (v.c. H7) and Co. Limerick (v.c. H8) (map, p.229).

Native of Arctic Eurasia and North America, and in the mountains of the Northern Hemisphere; in Europe on mountains south to C. Greece.

Variation in morphology and physiological ecology between Arctic and Alpine populations, all intolerant of high summer temperatures, was investigated by Mooney & Billings (1960). The germination ecology is described by Humlum (1980) and taxonomic variation is described by Chrtek & Sourkova (1992).

The reniform leaves, leafless inflorescence and winged, lenticular fruits distinguish this plant from all other species of Polygonaceae in Great Britain, Ireland and Europe.

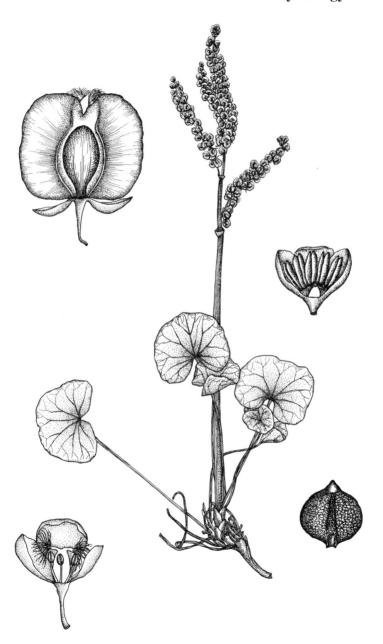

82 Emex spinosa (L.) Campd.   *Spiny Emex, Lesser Jack*

A glabrous, rather fleshy annual, prostrate to weakly erect. Leaves broadly ovate to oblong, obtuse, cuspidate, truncate or subcordate at the base, slightly fleshy. Male flowers in terminal and axillary pedunculate clusters. Female flowers axillary, sessile. Inner segments of fruiting perianth subacute; fruiting perianth 4–5 × 3–4 mm, including spines on the 3 outer tepals; spines 1-3 mm long, recurved. Nut trigonous, glabrous, brown, enclosed within the perianth. $2n = 20$ (Jaretzky 1927).

Native of S.W. Asia and warmer parts of the Mediterranean region; introduced as a weed in Australia and elsewhere. In Britain a rare casual with wool-shoddy or grain, formerly found in arable fields treated with shoddy, at docks and near mills and breweries.

83 Emex australis Steinh.   *Spiny Emex, Cathead*

Similar to *E. spinosa* but distinguished by its larger fruiting structures. Inner segments of fruiting perianth somewhat rounded with a terminal spiny arista. Fruiting perianth 4–8 × 10–13 mm, including spines on the 3 outer tepals; spines 5–6 mm long, spreading. $2n = 20$ (Jaretzky 1927).

Native of Southern Africa, widely introduced elsewhere, especially Australia, where the hybrid with 82 *E. spinosa* occurs (Putievsky, Weiss & Marshall 1980). Introduced to Britain as a very rare casual associated with wool-shoddy.

Emex spinosa 82 (A)
E. australis 83 (B)

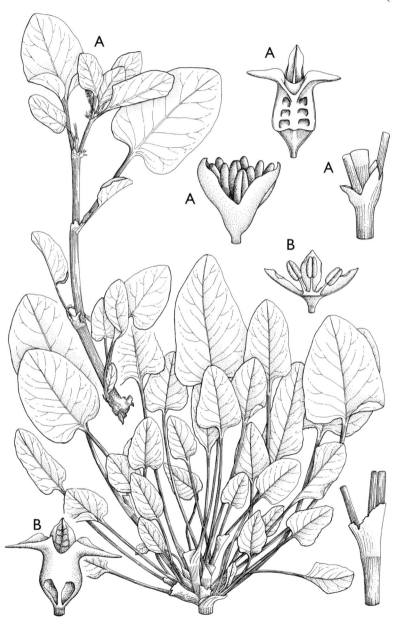

233

GLOSSARY

Acuminate: with a long fine point.

Acute: narrowed into a short point.

Acyclic: arranged spirally; not in whorls.

Amplexicaul: clasping the stem.

Androecium: the stamens collectively.

Apiculate: ending abruptly in a short point.

Appressed: lying close to the surface.

Arcuate: curved like a bow.

Arista: awn.

Ascending: sloping or curving upwards.

Asperous: rough to the touch.

Attenuate: gradually tapering.

Auriculate: furnished with ear-like appendages.

Biconvex: convex on two sides.

Caducous: dropping off early.

Caespitose: tufted.

Calyx: the sepals as a whole.

Capitat: with a knob-like head or tip.

Cauline (leaves): borne on the aerial part of the stem, especially the upper part, but not subtending a flower or inflorescence.

Ciliate: with regularly spaced hairs projecting from the margin.

Compressed: flattened.

Concolorous: of one colour throughout.

Confluent: running together.

Constricted: drawn together, narrowed in the middle.

Convex : having a more or less rounded surface.

Cordate: heart-shaped.

Coriaceous: leathery.

Corymb: a raceme (*q.v.*) with the pedicels becoming shorter towards the top, so that all the flowers are at approximately the same level.

Crenulate: having small rounded teeth.

Crisped: curled somewhat irregularly; unkempt.

Cuneate: wedge-shaped.

Cuneiform: wedge-shaped with the thin end at the base.

Cuspidate: tapering gradually into a rigid point.

Cyclic: arranged in whorls.

Decumbent (stems): lying on the ground but tending to rise at the end.

Deltate (in a plane shape): triangular.

Deltoid (in a solid object): shaped like the Greek letter Δ (delta).

Dentate: having sharp teeth with concave or straight edges.

Denticulate: with very small teeth.

Dioecious: having the sexes on different plants (*cf.* monoecious)

Distant: when similar parts are not close to each other.

Divaricate: spreading at a wide angle.

Eglandular: without glands.

Ellipsoid: elliptic in longitudinal section (of a solid object)

Elongate: drawn out, lengthened.

Entire: not toothed or dissected.

Excurrent: running beyond the margin, e.g veins of a leaf.

Exserted: protruding.

Fasciculate: clustered, growing in bundles.

Fastigiate: with branches clustered, parallel and erect, giving a narrow, elongated appearance.

Filiform: thread-like.

Fimbriate: with the margin divided into a fringe.

Flexuous: wavy.

Fruit (in *Rumex*): The nut together with the enveloping valves, the receptacle and the pedicel down to the point where it breaks away from the parent plant.

Fugacious: withering very rapidly.

Fusiform: spindle-shaped.

Glabrescent: becoming glabrous.

Glabrous: without hairs.

Gland: a small oblong or globular spot containing oil, resin, or other liquid; it can be sunk into the surface of, or protrude from, any part of a plant.

Glaucous: Bluish-grey.

Globose: spherical.

Gynoecium: the ovary, style(s) and stigma(ta) collectively; pistil(s).

Hastate: with equal, ± triangular basal lobes which are directed outwards.

Hermaphrodite: having the sexes on the same plant (*cf.* monoecious).

Heterophyllous: having leaves of more than one form or size.

Heterostylous: having styles of different lengths in individuals of the same species.

Homogenous: uniform in structure.

Hyaline: colourless and ±transparent or translucent.

Incumbent: overlapping.

Inflorescence: flowering branch or portion of the stem above the uppermost stem leaves, including its branches, bracts and flowers.

Infructescence: the mature inflorescence (*q.v.*) after seed set, with persistent fruits (*q.v.*), brown, reddish or yellow in colour.

Internode: the length of stem between two adjacent lobes.

Interrupted: not continuous.

Lacerate: deeply and irregularly divided and appearing as if torn.

Lacinia: a slender lobe.

Lanate: with woolly white hairs

Lanceolate: shaped like a lance- or spear-head; proportionately narrower than ovate (*q.v.*).

Lenticular: convex on both faces and ±circular in outline.

Ligulate: strap-shaped.

Linear: narrow, with the opposite margins parallel.

Lingulate: tongue-shaped.

Membranous: thin, dry and flexible, not green.

Midrib: median vein.

Monoecious: having the sexes in different flowers on the same plant.

Muricate: rough with short firm stubby projections.

Naked: lacking hairs or scales.

Nerve: a strand of strengthening and conductive tissue running through a leaf or modified leaf.

Node: a point on the stem from which one or more leaves arise.

Nut: a 1-seeded fruit with a woody pericarp (seed coat).

Obovate: broadest above the middle and ovate.

Obscure: indistinct.

Obtuse: blunt, apex less than 90°.

Ochrea, ocrea: sheathing organ formed by fusion of two stipules, arising from the base of the petiole and embracing the stem (characteristic of Polygonaceae).

Ovary: the female organ containing the ovules.

Ovate: shaped like the longitudinal section of a hen's egg.

Ovule: a structure containing the egg and developing into an embryo within the nut after fertilisation.

Palmate: lobed or divided like an outspread hand.

Panduriform: shaped like the sound box of a violin.

Panicle: strictly a branched racemose inflorescence, though often loosely applied to any branched inflorescence.

Papillae: small elongated projections.

Pedicel: the stalk of a single flower.

Pentamerous: with parts in fives.

Perianth: the floral envelope as a whole, including sepals and petals (where present).

Perianth-segments: The separate component parts of the perianth, especially when petals and sepals cannot be distinguished.

Persistent: not falling off.

Petaloid: brightly coloured and resembling petals.

Petiole: the stalk of a leaf.

Pilose: hairy, with rather long soft hairs.

Plano-convex: with one surface plane and the other convex.

Polygamous: having male, female and hermaphrodite flowers on the same or different plants.

Pubescent: shortly and softly hairy.

Punctate: dotted or shallowly pitted, often with glands.

Radical (of leaves): arising from the base of the stem or from the tip of a rhizome.

Reniform: kidney-shaped.

Reticulate: marked with a network, usually of veins.

Revolute: rolled downwards.

Rhizome: an underground stem living more than one growing season.

Rugose: wrinkled.

Sagittate: shaped like an arrow-head.

Scabrid: rough to the touch on account of numerous minute projections.

Scarious: thin, dry, membranous, not green and usually colourless.

Serrate: toothed like a saw with sharp teeth pointing forwards.

Sessile: without a stalk.

Simple: undivided, unbranched, not compound.

Sinuate: having a wavy outline.

Spathulate: paddle- or spatula-shaped.

Spiciform: resembling a spike.

Stamen: one of the male reproductive organs of a plant.

Striate: having fine lines, ridges or grooves.

Strigose: with stiff appressed hairs.

Style: the part of the gynoecium connecting the ovary with the stigma.

Sub (prefix): almost, nearly, somewhat.

Syncarpous (ovary): having the carpels united with each other.

Tepals: inner and outer perianth-segments.

Terminal: borne at the end of a stem and limiting its growth.

Tomentose: covered with short dense interwoven hairs.

Trigonous: 3-sided.

Triquetrous: 3-sided with sharp angles.

Truncate: ending abruptly, or (of a leaf-base) straight.

Tubercle: a ±spherical or ovoid swelling.

Undulate: wavy in a plane at right angles to the surface.

Unilocular: having a single cavity.

Valve (*Rumex*): inner 3 tepals that have reached maximum development following fertilization.

Whorl: the arrangement of organs in a circle round an axis.

Winged: furnished with an expansion or flange, as on a stem or petiole, etc.

REFERENCES

AKEROYD. J.R. 1986. *Polygonum* L. In A. STRID (ed.), *Mountain Flora of Greece*, **1**: 59–65. Cambridge.

AKEROYD. J.R. 1989. *Polygonum* Linnaeus. In S.M. WALTERS *et al.* (eds), *The European Garden Flora*, **3**: 125–128. Cambridge.

AKEROYD, J.R. 1993a. The distribution and status of *Rumex pulcher* L. in Ireland. *Irish Naturalists' Journal*, **24**: 284–285.

AKEROYD, J.R. 1993b. Give a dock a bad name. *Convivium*, **1** (**4**): 44–48.

AKEROYD, J.R. 2013. New nomenclatural combinations in *Persicaria* Miller and a new hybrid name in *Rumex* L. (Polygonaceae*). Contribuţii Botanice, Cluj*, **48**: 15–21.

AKEROYD, J.R. & BRIGGS, D., 1983a. Genecological studies of *Rumex crispus* L. I. Garden experiments using transplanted material. *New Phytologist*, 94: 309–323.

AKEROYD, J.R. & BRIGGS, D., 1983a. Genecological studies of *Rumex crispus* L. II. Variation in plants grown from wild-collected seed. *New Phytologist*, 94: 325–343.

AKEROYD, J.R. & RUTHERFORD, R.W. 1987. A small, knotty problem – what is *Polygonum equisetiforme* of gardens? *The Plantsman*, 8: 229–232.

AKEROYD, J.[R.], WOOLSTENHOLME, L. & POOLE, J. 2011. *Supplement to The wild plants of Sherkin, Cape Clear and adjacent islands of West Cork*. Sherkin Island Marine Station.

AL-BERMANI, A.-K.K.A., AL-SHAMMARY, K.I.A., BAILEY, J.P. & GORNALL, R.J. 1993. Contributions to a cytological catalogue of the British and Irish flora, 3. *Watsonia*, **19**: 269–271.

ARMITAGE, J.D. 2013. New combinations in *Persicaria amplexicaulis* (D. Don) Ronse Decr. and the reinstatement of the cultivar name 'Arun Gem'. *Hanburyana*, **7**: 47–50.

BAILEY, J.P. 1988. Putative *Reynoutria japonica* Houtt. × *Fallopia baldschuanica* (Regel) Holub hybrids discovered in Britain. *Watsonia*, **17**: 163–164.

BAILEY, J.P. 2001. *Fallopia × conollyana* The Railway-yard Knotweed. *Watsonia*, **23**: 539–541.

BAILEY, J.P. 2003. Japanese Knotweed *s.l.* at home and abroad. In L. CHILD, J.H. BROCK, K. PRACH, P. PYSEK, P.M. WADE & M. WILLIAMSON (eds), *Plant Invasions – ecological threats and management solutions*, 183–196. Leiden.

BAILEY, J.P., CHILD, L.E. & CONOLLY, A.P. 1996. A survey of the distribution of *Fallopia × bohemica* (Chrtek & Chrteková) J. Bailey (Polygonaceae) in the British Isles. *Watsonia*, **21**: 187–198.

BAILEY, J.P. & CONOLLY, A.P. 1991. Alien species of *Polygonum* and *Reynoutria* in Cornwall 1989–1990. *Botanical Cornwall Newsletter*, **5**: 3–46.

BAILEY, J.P. & CONOLLY, A.P. 2000. Prize-winners to Pariahs – a history of Japanese Knotweed s.l. in the British Isles. *Watsonia*, **23**: 93–110.

BAILEY, J.P. & SPENCER, M. 2003. New records for *Fallopia × conollyana*: is it truly such a rarity? *Watsonia*, **24**: 452–453.

BAILEY, J.P. & STACE, C.A. 1992. Chromosome number, morphology, pairing, and DNA values of species and hybrids in the genus *Fallopia* (Polygonaceae). *Plant Systematics and Evolution*, **180**: 29–52.

HALL, P.M. (ed., BEC Editorial Committee) 1937. 618/30. *Rumex magellanicus* Griseb. *Report of the Botanical Society and Exchange Club of the British Isles*, **11**: 228.

BEEFTINK, W. 1964. *Polygonum maritimum* L. in Nederland. *Gorteria*, **2**: 13–20.

BEERLING, D.J., BAILEY, J.P. & CONOLLY, A.P. 1994. *Fallopia japonica* (Houtt.) Ronse Decraene (*Reynoutria japonica* Houtt.; *Polygonum cuspidatum* Sieb. & Zucc.). (Biological Flora of the British Isles). *Journal of Ecology*, **82**: 959–979.

BRENAN, J.P.M. & SIMPSON, N.D. 1949. The results of two botanical journeys in Ireland in 1938–39. *Proceedings of the Royal Irish Academy*, 52 (B): 57–84.

BRITTON, C.E. 1933. British Polygona, Section *Persicaria*. *Journal of Botany, London*, **71**: 90–98.

BURTT, B.L. 1950. *Koenigia islandica* L. in Britain. *Kew Bulletin*, 5: 266.

CAMPDERÁ, F. 1819. *Monographie des* Rumex. Paris.

CAMUS, G.G. 1904. Renseignements bibliographiques sur les hybrides du genre *Rumex*. *Bulletin Herbarium Boissier*, sér.2, 4: 1232–1240.

CARLQUIST, S. 2003. Wood anatomy of Polygonaceae: analysis of a family with exceptional wood diversity. *Botanical Journal of the Linnean Society*, 141: 25–51.

CASTRO, D. DE & FONTES, F.C. 1946. Primeiro contacto citólogico con a flora halófilo dos salgados de Sacavém. *Broteria*, **15**: 38–46.

CAVERS, P.B. & HARPER, J.L. 1964. *Rumex obtusifolius* L. and *R. crispus* L. (Biological Flora of the British Isles). *Journal of Ecology*, **52**: 737–746.

CAVERS, P.B. & HARPER, J.L. 1967a. Studies in the dynamics of plant populations. I. The fate of seed and transplants introduced into various habitats. *Journal of Ecology*, **55**: 59–71.

CAVERS, P.B. & HARPER, J.L. 1967b. The comparative biology of closely related species living in the same area. IX. *Rumex*: the nature of adaptation to a sea-shore habitat. *Journal of Ecology*, **55**: 73–82.

CHEN, Q.-F. 1999 A study of resources of *Fagopyrum* (Polygonaceae) native to China. *Botanical Journal of the Linnean Society*, **130**: 53–64.

CHRTEK, J. & SOURKOVA, M. 1992. Variation in *Oxyria digyna*. *Preslia*, **64**: 207–210.

CLEMENT, E.J. 1983. Some rare alien Polygonaceae. *BSBI News*, No. 34: 29–31.

CONOLLY, A.P. 1977. The distribution and history in the British Isles of some alien species of *Polygonum* and *Reynoutria*. *Watsonia*, **11**: 291–311.

CONOLLY, A.P. 1991. *Polygonum lichiangense* W. Smith: rejected as a naturalized British species. *Watsonia*, **18**: 351–358.

CONSAUL, L.J., WARWICK, S.I. & McNEILL, J. 1991. Allozyme variation in the *Polygonum lapathifolium* L. complex. *Canadian Journal of Botany*, **69**: 2261–2270.

CURTIS, T.G.F. 1993. *Polygonum viviparum* in Ireland with particular reference to the flora and vegetation of the Mount Brandon Range, Co. Kerry. *Irish Naturalists' Journal*, **24**: 274–280.

DANIELS, R.E., McDONNELL, E.J. & RAYBOULD, A.F. 1998. The current status of *Rumex rupestris* Le Gall (Polygonaceae) in England and Wales, and threats to its survival and genetic diversity. *Watsonia*, **22**: 33–39.

DANSER, B.H. 1916. Bijlage 3. Mededelingen, gehouden door den heer Danser [*Rumex*-bastaarden]. *Nederlandsch Kruidkundig Archief*, **1915**: 103–116.

DANSER, B.H. 1922a. De Nederlandsche *Polygonum*-bastaarden. *Nederlandsch Kruidkundig Archief*, **1921**: 156–166.

DANSER, B.H. 1922b. Bijlage tot de Kennis der Neerlandsche *Rumex*. *Nederlandsch Kruidkundig Archief*, **1921**: 167–228.

DANSER, B.H. 1922c–1924. De Nederlandsche *Rumex*-bastaarden, I–III. *Nederlandsch Kruidkundig Archief*, **1921**: 229–265; **1922**: 175–210; **1923**: 232–270.

DANSER, B.H. 1922d. Fünf neue *Rumex*-Bastarde. *Recuil des Travaux botaniques néerlandais*, 19: 293–295.

DANSER, B.H. 1926. Beitrag zur Kenntnis der Gattung *Rumex*. *Nederlandsch Kruidkundig Archief*, **1925**: 414–484.

DANSER, B.H. 1927. Die Polygonaceen Niederlaendisch-Ostindiens. *Bulletin du Jardin Botanique de Buitenzorg*, ser.3, **8**: 119–261.

DARLINGTON, H.T. & STEINBAUER, G.P. 1961. The eighty-year period for Dr Beal's seed viability experiment. *American Journal of Botany*, **48**: 321–325.

DEGRAEVE, N. 1975a. Contribution a l'étude cytotaxonomique des *Rumex*, 1. Le genre *Rumex. Caryologia*, **28**: 187–201.

DEGRAEVE, N. 1975b. Contribution a l'étude cytotaxonomique des *Rumex*, 2. Le genre *Bucephalophora* Pau. *Caryologia*, **28**: 203–206.

DEGRAEVE, N. 1976. Contribution a l'étude cytotaxonomique des *Rumex*, 4. Le genre *Acetosa* Mill. *La Cellule*, **71**: 231–250.

DEMPSEY, R.E., GORNALL, R.J. & BAILEY, J.P. 1994. Contributions to a cytological catalogue of the British and Irish flora, 4. *Watsonia*, **20**: 63–66.

DOIDA, Y. 1960. Cytological studies in *Polygonum* and related genera, 1. *Botanical Magazine, Tokyo*, **73**: 337–340.

DOIDA, Y. 1961. Cytological studies in the genus *Polygonum*, 2. *Annual Report of the National Institute of Genetics, Japan*, **11**: 65.

DOIDA, Y. 1962. Consideration on the intrageneric differentiation in *Polygonum*. *Journal of Japanese Botany*, **37**: 3–12.

DONALDSON, F., DONALDSON, F., & McMILLAN, N.F. 1978. *Polygonum sagittatum* L.: its status in Co. Kerry. *Irish Naturalists' Journal*, **19**: 168.

DRUCE, G.C. 1913. 2184 (3). *Polygonum calcatum* Lindman. *Report of the Botanical Society and Exchange Club of the British Isles*, **3**: 176.

DRUCE, G.C. 1924. 2209 (2). *R. arifolius* All., 2210 (13). *R. obovatus* Danser. *Report of the Botanical Society and Exchange Club of the British Isles*, **7**: 58–62.

DRUCE, G.C. (1930). 618/13 *Rumex maritimus* x *palustris* = R. x *henrardii* Danser. *Report of the Botanical Society and Exchange Club of the British Isles*, **9**: 36.

EDMAN, G. 1929. Zur Entwicklungsgeschichte der Gattung *Oxyria* Hill, nebst zytologischen, embryologischen und systematischen Bemerkungen über einige andere Polygonaceen. *Acta Horti Bergiani*, **9**: 165–291.

EKMAN, S. & KNUTTSON, T. 1994. Nomenclatural changes in *Persicaria*. *Nordic Journal of Botany*, **14**: 23–25.

ENGELL, K. 1973. A preliminary morphological, cytological, and embryological investigation in *Polygonum viviparum*. *Botanisk Tidsskrift*, **67**: 305–316.

ENGELL, K. 1978. Morphology and cytology of *Polygonum viviparum* in Europe, 1. The Faroe Islands. *Botanisk Tidsskrift*, **72**: 113–118.

FERGUSON, I.K. & FERGUSON, L.F. 1974. *Polygonum maritimum* L. new to Ireland. *Irish Naturalists' Journal*, **18**: 95.

FERNANDES, A. 1983. Sur l'existence de formes octoploïdes chez l'agregat du *Rumex acetosella* dans la Péninsule Ibérique. *Revista de Biologia*, **12**: 341–362.

FLOVIK, K. 1940. Chromosome numbers and polyploidy within the flora of Spitzbergen. *Hereditas*, **26**: 430–444.

FOUST, C.M. 1992. *Rhubarb. The wondrous drug*. Princeton.

FOUST, C.M. & MARSHALL, D.E. 1991. Culinary rhubarb production in North America: history and recent statistics. *HortScience*, **26**: 1360–1363.

FREEMAN, C.C. & REVEAL, J.L. 2005. Polygonaceae, in Flora of North America Editorial Committee (eds), *Flora of North America*, **5**: 216–601. New York & Oxford.

GILBERT, O. 1992. The ecology of an urban river. *British Wildlife*, **3**: 129–136.

GLOB, P.V. 1969. *The bog people*. London.

GREEN, P.R., GREEN, I.P. & CROUCH, G.A. 1997. *The Atlas Flora of Somerset*. Yeovil.

GRIGSON, G. 1955. *The Englishman's Flora*. London.

HACKNEY, P. (ed.) 1992. *Stewart and Corry's Flora of the North-east of Ireland*. 3rd ed., Belfast: Institute of Irish Studies.

HAGERUP, O. 1926. Konsdelenes bygning og udvikling hos *Koenigia islandica* L. *Meddr Grønland*, **58**: 199–204.

HAMET-AHTI, L. & VIRRANSKO, V. 1970. Chromosome numbers of some vascular plants of north Finland. *Anales Botanici Fennici*, **7**: 177–181.

HANSON, G. 2002. Yet more *Persicaria capitata*. *BSBI News*, No. 98: 52.

HARALDSON, K. 1978. Anatomy and taxonomy in Polygonaceae subfam. Polygonoideae Meisn. emend. Jaretzky. *Symbolae Botanicae Upsallensis*, **22**: 1–93.

HARRIS, W. 1970. Genecological aspects of flowering and vegetative reproduction in *Rumex acetosella* L. *New Zealand Journal of Botany*, **8**: 99–113.

HART, M.L., BAILEY, J.P., HOLLINGSWORTH, P.M. & WATSON, K.J. 1997. Sterile species and fertile hybrids of Japanese knotweeds along the River Kelvin. *The Glasgow Naturalist*, **23** (**2**): 18–22.

HAYWARD, I.M. & DRUCE, G.C. 1919. Polygonaceae. In *The adventive Flora of Tweedside*, 206–212. Arbroath.

HEDBERG, O. 1997. The genus *Koenigia* L. emend. Hedberg (Polygonaceae). *Botanical Journal of the Linnean Society*, **124**: 295–330.

HICKMAN, J.C. 1974. Pollination by ants: a low-energy system. Science **184**: 1290–1292.

HOLLINGSWORTH, M.L. & BAILEY, J.P. 2000. Hybridisation and clonal diversity in some introduced *Fallopia* species. *Watsonia*, **23**: 111–121.

HOLYOAK, D.T. 1995. *Rumex frutescens* Thouars × *R. obtusifolius* L. (Polygonaceae), a previously undescribed hybrid dock, and new records of *R.* × *wrightii* Lousley in West Cornwall (v.c. 1). *Watsonia*, **20**: 412–415.

HOLYOAK, D.T. 2000. Hybridization between *Rumex rupestris* Le Gall (Polygonaceae) and other docks. *Watsonia*, **23**: 83–92.

HONG, S.-P. 1992. Taxonomy of the genus *Aconogonon* (Polygonaceae) in Himalaya and adjacent regions. *Symbolae Botanicae Upsallensis*, **30** (**2**): 11–118.

HONG, S.-P. 1993. Reconsideration of the generic status of *Rubrivena* (Polygonaceae, Persicarieae). *Plant Systematics and Evolution*, **186**: 95–122.

HOOKER, J.D. 1881. *Polygonum sacchalinense*. *Curtis's Botanical Magazine*, **107**: t. 6540.

HULL, P. & NICHOLL, M.J. 1982. Hybridization between *Rumex aquaticus* L. and *Rumex obtusifolius* L. in Britain. *Annals of Botany*, **49**: 127–129.

HUMLUM, C. 1980. Germination ecology in the arctic-alpine species *Oxyria digyna* and the alpine species *Oxyria elatior*. *Botanisk Tidsskrift*, **75**: 173–180.

HYLANDER, N. 1966. Polygonaceae. In *Nordisk Kärlväxtflora*, **2**: 329–385.

ICHIKAWA, S. *et al.* 1971. Chromosome number, volume and nuclear volume relationships in a polyploid series ($2x–20x$) of the genus *Rumex*. *Canadian Journal of Genetics and Cytology*, **13**: 842–863.

IDLE, E.T. 1968. *Rumex aquaticus* L. at Loch Lomondside. *Transactions and Proceedings of the Botanical Society of Edinburgh*, **40**: 445–449.

JALAS, J. & LINDHOLM, R.K. 1975. Biosystematics of *Rumex longifolius* DC. of Fennoscandia and the Pyrenees. *Anales del Instituto Botánico A.J. Cavanilles*, **32**: 197–202.

JALAS, J. & SUOMINEN, J. 1979. *Atlas Florae Europaeae*, **4**. *Polygonaceae*. Helsinki: Societas Biologica Fennica Vanamo.

JARETZKY, R. 1927. Einige Chromosomenzahlen aus der Familie der Polygonaceae. *Berichte der Deutschen Botanischen Gesellschaft*, **45**: 48–54.

242

JARETZKY, R. 1928. Histologische und karyologische Studien an Polygonaceen. *Jahrbuch wiss. Bot.*, **69**: 357–490.

JERMYN, S.T. 1975. *Flora of Essex*. Colchester.

JEX-BLAKE, Lady M. 1948. *Some Wild Flowers of Kenya*. Nairobi.

JOHNSON, L.A.S. & BRIGGS, B.L. 1962. Taxonomic and cytological notes on *Acetosa* ansd *Acetosella* in Australia. *Contributions from the N.S.W. national Herbarium*, **3**: 165–169.

KARLSSON, T. (ed.) 2000. 35. Polygonaceae. In B. Jonsell (ed.), *Flora Nordica*, **1**, *Lycopodiaceae to Polygonaceae*, pp. 235–318. Stockholm: Bergius Foundation.

KENT, D.H. 1975. *The historical Flora of Middlesex*. London: Ray Society.

KENT, D.H. 1977. *Rumex* × *lousleyi* hybr. nov. (*R. cristatus* DC. × *R. obtusifolius* L.). *Watsonia*, **11**: 313–314.

KIHARA, H. 1927. Uber die Vorbehandlung einiger pflanzlicher Objekte bei dr Fixierung der Pollenmutterzellen. *Botanical Magazine, Tokyo*, 41: 124–128.

KIHARA, H. 1927. Chromosomenzahlen und systematische Gruppierung der *Rumex*-Arten. *Zeitschrift fur Zellforschung und mikroskopische Anatomie*, **4**: 475–481.

KIM, M.-H., PARK, J.-H. & PARK, C.-W. 2000. Flavonoid chemistry of *Fallopia* section *Fallopia* (Polygonaceae). *Biochemical systematics and Ecology*, **28**: 433–441.

KIM, S.-T., KIM, M.-H. & PARK, C.-W. 2000. A systematic study of *Fallopia* section *Fallopia* (Polygonaceae). *Korean Journal of Plant Taxonomy*, **30**: 35–54.

KIMURA, Y., KOZOWA, M., BABA, K. & HATA, K. 1983. New constituents of roots of *Polygonum cuspidatum*. *Planta Medica*, **48**: 164–168.

KITCHENER, G.D. 1996a. Hybrid docks and willowherbs in Counties Sligo, Leitrim and Mayo. *Irish Botanical News*, **6**: 23–25.

KITCHENER, G.D. 1996b. *Rumex* × *fallacinus* Haussknecht – a hybrid dock new to the British Isles. *Transactions of the Kent Field Club*, **15**: 39–40.

KITCHENER, G.D. 2002. *Rumex* × *xenogenus* (Rech. fil.) (Polygonaceae), the hybrid between Greek and Patience Docks, found in Britain. *Watsonia*, **24**: 209–213.

KUROSAWA, S. 1971. Cytological studies on some eastern Himalayan plants and their related species. *Bulletin of the University Museum Tokyo*, **2**: 355–364.

LAW, R., COOK, R.E.D. & MANLOVE, R.J. 1983. The ecology of flower and bulbil production in *Polygonum viviparum*. *Nordic Journal of Botany*, **3**:559–565.

LE DEUNFF, Y. 1974. Hétérogénéité de la germination des semences de *Rumex crispus* mise en evidence et essai d'interprétation. *Physiologia Plantarum*, **32**: 342–346.

LEES, F.A., (C.A. CHEETHAM & W.A. SLEDGE, eds). 1942. *A Supplement to the Yorkshire Floras*. London.

LEWIS, W.H. & ELVIN-LEWIS, M.P.F. 2003. *Medical Botany*. 2nd ed. NJ & Canada.

LI, ANJEN *et mult. al.* 2003. Polygonaceae, in *Flora of China* 5: 277–350. [http://flora.huh.harvard.edu/china/mss/volume05/Polygonaceae.pdf]

LINDLEY, J. 1846. *Fagopyrum cymosum*. *Botanical Register* **1846**: t.26.

LINDMAN, C.A.M. 1912. Wie ist die Kollektivart *Polygonum aviculare* zu spalten? *Svensk Botanisk Tidskrift*, **6**: 673–696.

LOUSLEY, J.E. 1939a. *Rumex aquaticus* as a British plant. *Journal of Botany, London*, **77**: 149–152.

LOUSLEY, J.E. 1939b. Notes on British Rumices, 1. *Report of the Botanical Society and Exchange Club of the British Isles*, **12**: 118–157.

LOUSLEY, J.E. 1944. Notes on British Rumices, 2. *Report of the Botanical Society and Exchange Club of the British Isles*, **12**: 547–585.

LOUSLEY, J.E. 1950. 615/25. *Polygonum cognatum* Meisner. *Watsonia*, **1**: 319–320.

LOUSLEY, J.E. 1953a. *Rumex cuneifolius* and a new hybrid. *Watsonia*, **2**: 394–397.

LOUSLEY, J.E. 1953b. 615/25. *Polygonum cognatum* Meisner. *Watsonia*, **2**: 414.

LOUSLEY, J.E. 1955. *Polygonum senegalense* Meisn. *Proceedings of the Botanical Society of the British Isles*, 1: 493–494.

LOUSLEY, J.E. 1967. 325/p. *Rumex pseudonatronatus* (Borb.) Murb. *Proceedings of the Botanical Society of the British Isles*, **7**: 25–26.

LOUSLEY, J.E. 1969. 325/p. *Rumex pseudonatronatus* (Borb.) Murb. *Proceedings of the Botanical Society of the British Isles*, **7**: 561.

LOUSLEY, J.E. 1976. Three species of Polygonaceae established in Britain. *Watsonia*, **11**: 144–146.

LOUSLEY, J.E. & WILLIAMS, J.T. 1975. *Rumex* L. In C.A. STACE (ed.), *Hybridization and the flora of the British Isles*: 278–292. London.

LÖVE, Á. 1940. Polyploidy in *Rumex acetosella* L. *Nature, London*, **145**: 351.

LÖVE, Á. 1967. IOPB Chromosome Number Reports, 13. *Taxon*, **16**: 445–461.

LÖVE, Á. 1968. IOPB Chromosome Number Reports, 19. *Taxon*, **17**: 573–577.

LÖVE, Á. 1983. The taxonomy of *Acetosella*. *Botanica Helvetica*, **93**: 145–168.

LÖVE, Á. 1986. IOPB Chromosome Number Reports, 40. *Taxon*, **35**: 195–198.

LÖVE, Á. & LÖVE, D. 1948. *Chromosome numbers of northern plant species*. Reykjavik.

LÖVE, Á. & LÖVE, D. 1956. Chromosomes and taxonomy of eastern North American *Polygonum*. *Canadian Journal of Botany*, **34**: 501–521.

LÖVE, Á. & LÖVE, D. 1961. Chromosome numbers of Central and Northwest Europe plant species. *Opera Botanica*, **5**: 1–581.

LÖVE, Á. & LÖVE, D. 1982. IOPB Chromosome Number Reports, 74. *Taxon*, **31**: 120–126.

LUSBY, P.S. 1999. *Koenigia islandica* L. (Polygonaceae). In M.J. Wigginton (ed.), *British Red Data Books.1. Vascular Plants*, 3rd ed: 206. Peterborough: Joint Nature Conservation Committee.

MABEY, R. 1996. *Flora Britannica*. London.

MALLICK, R. 1968. Cytotaxonomy of some Indian Polygonaceae. *Nucleus (Calcutta)*, **1968**, Suppl.: 36–38.

MARSHALL, D.E. 1988. A Bibliography of Rhubarb and *Rheum* species. *USDA Bibliography of Literature and Agriculture*, 62.

MAUN, M.A. 1974. Reproductive biology of *Rumex crispus*. Phenology, surface area of chlorophyll-containing tissue, and contribution of the perianth to reproduction. *Canadian Journal of Botany*, **52**: 2181–2187.

MEISSNER, C.F. 1826. *Monographiae Generis Polygoni Prodromus*. Geneva.

MEISSNER, C.F. 1856. Polygonaceae Subordo 2. Polygonea, *et seq*. In A.P. De Candolle, *Prodromus Systematis naturalis Regni vegetabilis*, **14**: 28–186. Paris.

MEERTS, P. BRIANE, J.-P. & LEFEBVRE, C. 1990. A numerical taxonomic study of the *Polygonum aviculare* complex (Polygonaceae) in Belgium. *Plant Systematics and Evolution*, **173**: 7189.

MERTENS, T.R. & RAVEN, P.H. 1965. Taxonomy of *Polygonum* section *Polygonum* (*Avicularia*) in North America. *Madroño*, **18**: 85–91.

MESICEK, J. & SOJÁK, J. 1973. Karyological and taxonomic observations on *Dracocephalum* Bunge and *Koenigia islandica* L. *Folia Geobotanica Phytotaxonomica, Praha*, **8**: 105–112.

MITCHELL, J. 1982. Two new Scottish dockens. *B.S.B.I. Scottish Newsletter*, **4**: 8–9.

MITCHELL, J. 1990. *Rumex* × *dumulosus* Hausskn. (*R. aquaticus* × *R. sanguineus*) a new Scottish hybrid. *B.S.B.I. Scottish Newsletter*, **12**: 23–24.

MONTGOMERY, L., KHALAF, M., BAILEY, J.P. & GORNALL, R.J. 1997. Contributions to a cytological catalogue of the British and Irish flora, 5. *Watsonia*, **20**: 365–368.

MOONEY, H.A. & BILLINGS, W.D. 1961. Comparative physiological ecology of Arctic and Alpine populations of *Oxyria digyna*. *Ecological Monographs*, **31**: 1–29.

MULLIGAN, G. 1957. Chromosome numbers of Canadian weeds, 1. *Canadian Journal of Botany*, **35**: 779–789.

MULLIGAN, G. 1959. Chromosome numbers of Canadian weeds, 2. *Canadian Journal of Botany*, **37**: 81–92.

MURBECK, S. 1899. Die nordeuropäischen Formen der Gattung *Rumex*. *Botaniska Notiser*, **1899**: 1–42.

MURBECK, S. 1913. Zur Kentniss der Gattung *Rumex*. *Botaniska Notiser*, **1913**: 201–237.

den NIJS, J.C.M. 1974. Biosystematic studies of the *Rumex acetosella* complex. I. Angiocarpy and chromosome numbers in France. *Acta Botanica Neerlandica*, **23**: 655–675.

den NIJS, J.C.M. 1976. Biosystematic studies of the *Rumex acetosella* complex. II. The Alpine region. *Acta Botanica Neerlandica*, **25**: 417–447.

den NIJS, J.C.M. 1984. Biosystematic studies of the *Rumex acetosella* complex (Polygonaceae). VIII. A taxonomic revision. *Feddes Repertorium*, **95**: 43–66.

NORDHAGEN, R. 1963. Studies on *Polygonum oxyspermum* Mey. et Bge., *Polygonum raii* Bab. and *P. raii* subsp. *norvegicum* Sam. *Norske Videnskaps-Akademi, Matematisk-Naturvidenskappelig Klasse, Avhandliger*, n.s., **5**: 1–40.

O'MAHONY, T. 2003. A new key to the native annual *Persicaria* (knotweed) species in Britain and Ireland, and an overlooked diagnostic character in *P. hydropiper* (L.) Spach (water-pepper). *Irish Botanical News*, 13: 15–18

PARNELL, J.A.N. & SIMPSON, D.A. 1989. Hybridization between *Polygonum mite* Schrank, *P. minus* Huds. and *P. hydropiper* L. in Northern Ireland with comments on theior distinctions. *Watsonia*, **17**: 265–272.

PARTRIDGE, J.W. 2001. *Persicaria amphibia* (L.) Gray (*Polygonum amphibium* L.) (Biological Flora of the British Isles). *Journal of Ecology*, 89: 487–501.

PÓLYA, L. 1950. Magyarországi novényfajok kromoszómaszámi, II (Chromosome numbers of Hungarian plants, 2). *Annales biologicae Universitatis Debreceniensis*, **1**: 46–65.

PRAEGER, R. Ll. 1942. Additional records for the flora of North-east of Ireland. *Irish Naturalists' Journal*, **8**: 35–36.

PRESS, J.R. 1988. Infraspecific variation in *Rumex bucephalophorus*. *Botanical Journal of the Linnean Society*, **97**: 344–355.

PROBATOVA, N.S. & SOKOLOVSKAYA, A.P. 1989. Числа хромосом сосудистых растений из Приморского края, Приамурья, Северной Корякии, Камчатки и Сахалина [Chromosome numbers in vascular plants from Primorye Territory, the Amur River basin, north Koryakia, Kamchatka and Sakhalin]. *Botanicheskii Zhurnal (St Petersburg)*, **73**: 290–293.

PUTIEVSKY, E., WEISS, P.W. & MARSHALL, D.R. 1980. Interspecific hybridization between *Emex australis* and *E. spinosa*. *Australian Journal of Botany*, **28**: 323–328.

RACKHAM, O. 1961. Ecological significance of hybridisation between *Rumex sanguineus* and *R. conglomeratus*. *Proceedings of the Botanical Society of the British Isles*, **4**: 332.

RAFFAELLI, M. 1979. Contributi alla conoscenza del genere *Polygonum* L. 2. *Polygonum bellardii* All. *Webbia*, **33**: 327–342.

RAFFAELLI, M. 1982. Contributi alla conoscenza del genere *Polygonum* L. 4. Le specie italiane della sect. *Polygonum*. *Webbia*, **35**: 361–406.

RATCLIFFE, D. 1959. The habitat of *Koenigia islandica* L. in Scotland. *Transactions and Proceedings of the Botanical Society of Edinburgh*, **37**: 272–275.

RAVEN, J.E. 1952. *Koenigia islandica* L. in Scotland. *Watsonia*, **2**: 188–190.

RECHINGER, K.H. 1932. Vorarbeiten zu einer Monographie der Gattung *Rumex*, I. *Beihefte zum Botanischen Centralblatt*, **49 (2)**: 1–132.

RECHINGER, K.H. 1933a. Vorarbeiten zu einer Monographie der Gattung *Rumex*, II. Die Arten der Subsektion *Patientiae*. *Reprium novarum Specierum Regni vegetabilis*, **31**: 225–283.

RECHINGER, K.H. 1933b. Vorarbeiten zu einer Monographie der Gattung *Rumex*, III. Die Sud- und Zentralamerikanischen Arten der Gattung *Rumex*. *Arkiv för Botanik*, Ser. 1, **26**: 1–58.

RECHINGER, K.H. 1935. Vorarbeiten zu einer Monographie der Gattung *Rumex*, IV. Die australischen und neuseeländischen Arten der Gattung *Rumex*. *Österreichische botanische Zetschrift*, **84**: 1–52.

RECHINGER, K.H. 1937. Vorarbeiten zu einer Monographie der Gattung *Rumex*, V. The North American species of *Rumex*. *Publications of the Field Museum of Natural History*, Bot. Ser., **17 (1)**: 1–151.

RECHINGER, K.H. 1939. Vorarbeiten zu einer Monographie der Gattung *Rumex*. VI. Versuch einer natürlichen Gliederung des Formenkreises von *Rumex bucephalophorus* L. *Botaniska Notiser*, **1939**: 485–504.

RECHINGER, K.H., 1948. Beiträge zur Kenntnis von *Rumex* Subgen. *Lapathum*. IX. *Candollea*, **11**: 229–241.

RECHINGER, K.H. 1949. Vorarbeiten zu einer Monographie der Gattung *Rumex*, VII. Rumices Asiatici. *Candollea*, **12**: 19–152.

RECHINGER, K.H. 1954. Vorarbeiten zu einer Monographie der Gattung *Rumex*, VIII. Monograph of the genus *Rumex* in Africa. *Botaniska Notiser Supplement*, **3** (3): 1–114.

RECHINGER, K.H., 1958a. *Rumex*. In G. Hegi, *Illustrierte Flora von Mittel-Europa*, ed.2, **3 (2)**: 353–400.

RECHINGER, K.H., 1961. Notes on *Rumex acetosa* L. in the British Isles (Beiträge zur Kenntnis von *Rumex*. XV). *Watsonia*, **5**: 64–66.

RECHINGER K.H., 1964. *Rumex* L. in T.G. TUTIN *et al.* (eds) *Flora Europaea*, 1, pp. 82–89. Cambridge.

RECHINGER, K.H. 1984. Vorarbeiten zu einer Monographie der Gattung *Rumex*, IX. *Rumex* (Polygonaceae) in Australia: a reconsideration. *Nuytsia, 5*: 75–122.

RECHINGER, K.H. (1990) *Rumex* subgen. *Rumex* sect. *Axillares* (Polygonaceae) in South America. *Plant Systematics and Evolution*, 172: 151–192.

REYNOLDS, S. 1998. Plant records from Co Limerick (H8) in 1997. *Irish Naturalists' Journal*, **26**: 129–132.

REYNOLDS, S. 2002. *A catalogue of alien plants in Ireland*. Occasional Papers No.14. National Botanic Gardens, Glasnevin, Dublin.

ROBERTS, R.H. 1977. *Polygonum minus* Huds. × *P. persicaria* L. in Anglesey. *Watsonia*, **11**: 255–256.

ROBERTY, G. & VAUTIER, S. 1964. Les genres de Polygonacées. *Boissiera*, **10**: 7–128.

RONSE DECRAENE, L.P. & AKEROYD, J.R. 1988. Generic limits in *Polygonum* and related genera (Polygonaceae) on the basis of floral characters. *Botanical Journal of the Linnean Society*, **98**: 321–371.

RONSE DECRAENE, L.P. 1989. The flower of *Koenigia islandica* L. (Polygonaceae): an interpretation. *Watsonia*, **17**: 419–423.

RUDYKA, E.G. 1995. Chromosome numbers in vascular plants from the southern part of the Russian Far East. *Botaniceskij Zurnal (St Petersburg)*, **80**: 87–90.

RUMSEY, F.J. 1999. *Rumex* × *akeroydii* – a new dock hybrid. *Watsonia*, **22**: 413–416.

SALT, D.T. & WHITTAKER, J.B. 1998. *Insects on dock plants*. Naturalists' Handbooks 26. The Richmond Publishing Co. Ltd., Slough.

SANDWITH, C.I. & SANDWITH, N.Y. 1936. 618/3 × 29 *Rumex crispus* L. × *obovatus* Danser. × *R. bontei* Danser. *Report of the Botanical Society and Exchange Club*, **11**: 40.

SARKAR, M.M. 1958. Cytotaxonomic studies on *Rumex* section *Axillares*. *Canadian Journal of Botany*, **36**: 947–966.

SCHMID, K. 1983. Untersuchungen an *Polygonum aviculare* s.l. in Bayern. *Mitteilungen der botanischen Staatssammlung München*, **19**: 29–149.

SCHOLZ, H. 1977. Bemerkungen zur Merkmalsgeographie des *Polygonum aviculare*, insbesondere des *P. arenastrum*. *Verhandlungen des botanisches Vereins der Provinz Brandenburg*, **113**: 13–22.

SCULLY, R.W. 1890. *Report of the Botanical Exchange Club of the British Isles*, **1889**: 267.

SELL, P.D. & AKEROYD, J.R. 1988. *Polygonum hydropiper* L. var. *densiflorum* A. Braun. *Watsonia*, **17**: 176–177.

SHARMA, A.K. & CHATTERJI, T. 1960. Chromosome studies of some members of the Polygonaceae. *Caryologia*, **13**: 486–506.

SHIMAMURA, T. 1929. Meiosis in *Rumex pulcher* L. *Journal of the Royal Microscopical Society*, **49**: 211–216.

SIMMONDS, N.W. 1945. *Polygonum persicaria* L., *P. lapathifolium* L. and *P. nodosum* Pers. (Biological Flora of the British Isles). *Journal of Ecology*, **33**: 117–143.

SKALINSKA, M. 1950. Studies in chromosome numbers of Polish Angiosperms. *Acta Societata Botanicorum Polonia*, **20 (1)**: 45–68.

SLEDGE, W.A. 1934. 615/7g. *Polygonum persicaria* L. subsp. *hirticaule* Danser; 615/35 *P. bungeanum* Turcz.; 615/36. *P. pennsylvanicum* L. *Report of the Botanical Society and Exchange Club of the British Isles*, x (iii): 481.

SMALL, J. 1895. A monograph of the North American species of the genus *Polygonum*. *Memoirs of Department of Botany, Columbia College*, **1**: 1–183.

SOJAK, J. 1974. Observations on the genus *Truellum* Houtt. (Polygonaceae). *Preslia*, **46**: 139–156.

SOKOLOVSKAYA, A.P. & STRELKOVA, A.S. 1938. Явление полиплоидии в высокогорьях Памира и Алтая [The phenomenon of polyploidy in the high mountains of Pamir and Altai]. *Doklady Akademii Nauk SSSR*, **21**: 68–71.

SÖYRINKI, N. 1989. Fruit production and seedlings in *Polygonum viviparum*. *Memoranda Societas pro Fauna et Flora Fennica*, **65**: 13–15.

SPENCER, K. 1987. *The magic of green buckwheat*. York.

STACE, C.A. *et al.* 1975 *Polygonum* L. In C.A. Stace (ed.), *Hybridization and the flora of the British Isles*, 273–278. London.

STACE, C.A. 2002. A new name in the British flora: *Persicaria* × *fennica* (Reiersen) Stace. *Watsonia*, **24**: 109–110.

STACE, C.A. 2010. *A new Flora of the British Isles*. 3rd ed. Cambridge.

STEARN, W.T. 1969. *Polygonum*. In P.M. SYNGE (ed.), *Supplement to the Dictionary of Gardening*, ed.2: 466–468. Oxford.

STEVENS, N.E. 1912. Observations on heterostylous plants. *Botanical Gazette*, **53**: 277–308.

STEWARD, A.N. 1930. The Polygonaceae of Eastern Asia. *Contributions from the Gray Herbarium, Harvard*, n.s. **5**, no.88: 1–129.

STYLES, B.T. 1962. The taxonomy of *Polygonum aviculare* and its allies in Britain. *Watsonia*, **5**: 177–214.

SUGIURA, T. 1931. A list of chromosome numbers in angiospermous plants. *Botanical Magazine, Tokyo*, **45**: 553–556.

SUKOPP, H. & SUKOPP, U. 1988. *Reynoutria japonica* Houtt. in Japan and Europe. *Veröffentlichungen der Geobotanisches Institut ETH, Stiftung Rübel, Zürich*, **98**: 354–372.

TIEBRE, M.-S. BIZOUX, J.-P., HARDY, O.J., BAILEY, J.P. & MAHY, G. 2007. Hybridization and morphogenetic variation in the invasive alien *Fallopia* (Polygonaceae) complex in Belgium. *American Journal of Botany*, **94**: 1900–1910.

TIMSON, J. 1963. The taxonomy of *Polygonum lapathifolium* L., *P. nodosum* Pers. and *P. tomentosum* Schrank. *Watsonia*, **5**: 386–395.

TIMSON, J. 1965. A study of hybridization in *Polygonum* section *Persicaria*. *Journal of the Linnean Society, Botany*, **59**: 155–161.

TIMSON, J. 1966. *Polygonum hydropiper* L. (Biological Flora of the British Isles). *Journal of Ecology*, **54**: 815–821.

TIMSON, J. 1975. *Polygonum* L. sect. *Persicaria* (Mill.) DC. In C.A. STACE (ed.), *Hybridization and the Flora of the British Isles*: 274–277. London.

TOOLE, E.H. & BROWN, E. 1946. The final results of the Duvel buried seed experiments. *Journal of Agricultural Research*, **72**: 201–210.

TRZCINSKA-TACIK, H. 1963. Studies on the distribution of synanthropic plants. 2. *Rumex confertus* Willd. in Poland. *Fragmenta Floristica et Geobotanica*, 9: 73–84.

TUTIN, T.G. *et al.* (eds) 1993. *Flora Europaea*, 1. 2nd ed. Cambridge: Cambridge.

WEBB, D.A. 1984. *Polygonum mite* Schrank in Ireland. *Irish Naturalists' Journal*, **21**: 283–286.

WEBB, D.A. & AKEROYD, J.R. 1991. Inconstancy of seashore plants. *Irish Naturalists' Journal*, **23**: 384–385.

WEISEL, Y. 1962. Ecotypic differentiation in the flora of Israel, 2. Chromosome counts in some ecotypic pairs. *Bulletin of the Research Council of Israel, section D, Botany*, **11**: 174–176.

WILLIAMS, J.T. 1971. Seed polymorphism and germination, 2. The role of hybridization in the germination polymorphism of *Rumex crispus* and *R. obtusifolius*. *Weed Research*, **11**: 12–21.

WISSKIRCHEN, R. 1991. Zur Biologie und Variabilitat von *Polygonum lapathifolium*. L. *Flora*, **185**: 267–295.

WISSKIRCHEN, R. 1995. Zur Bestimmung der Unterarten von *Polygonum lapathifolium*. L. s.l. *Floristische Rundbriefe*, 29 **29 (1)**: 1–25.

WOLF, S.J. & McNEILL, J. 1987. Cytotaxonomic studies on *Polygonum* section *Polygonum* in eastern Canada and the adjacent United States. *Canadian Journal of Botany*, **65**: 647–652.

WULFF, H.D. 1939. Chromosomenstudien an der schleswig-holsteinischen Angiospermen-Flora 4. *Berichte der Deutschen botanischen Gesellschaft*, **57**: 424–431.

YANG, J. & WANG, J.W. 1991. A taxometric analysis of characters of *Polygonum lapathifolium* L. *Acta Phytotaxonomica Sinica*, **29**: 258–263.

ZIBURSKI, A., KADEREIT, J.W. & LEINS. P. 1986. Quantitative aspects of the hybridization of mixed populations of *Rumex obtusifolius* L. and *R. crispus* L. (Polygonaceae). *Flora*, **178**: 233–242.

INDEX TO ENGLISH NAMES

INDEX TO SCIENTIFIC NAMES

Roman numbers refer to species numbers. **Accepted names are in bold font**, *synonyms in italics*. Hybrids are often mentioned under both parents, but the number refers to the species under which they are described and discussed.

FALLOPIA Adans.
Fallopia aubertii (Louis Henry) Holub - sub 40
Fallopia baldschuanica(Regel) Holub **40**
Fallopia × **bohemica** (Chrtek & Chrtková) J.P. Bailey **43**
Fallopia × **conollyana** J.P. Bailey - sub 43
Fallopia convolvulus (L.) Á. Löve **38**
F. convolvulus var. **subalata** (Lej. & Court.) D.H. Kent - sub 38
Fallopia dumetorum (L.) Holub **39**
Fallopia japonica (Houtt.) Ronse Decr. var. **japonica 41**
F. japonica var. **compacta** (Hook. f.) J.P. Bailey - sub 41
Fallopia sachalinensis (F. Schmidt ex Maxim.) Ronse Decr. **42**
Fallopia scandens (L.) Holub - sub 39

KOENIGIA L.
Koenigia islandica L. **22**

MUEHLENBECKIA Meisn.
Muehlenbeckia axillaris (Hook. f.) Endl. - sub 44
Muehlenbeckia complexa Meisn. **44**

OXYGONUM Burch. ex Campd.
Oxygonum sinuatum (Hochst. & Steud. ex Meisn.) Dammer **26**

OXYRIA Hill
Oxyria digyna (L.) Hill **81**

PERSICARIA (L.) Mill.
Persicaria affinis (D. Don) Ronse Decr. = sub 8
Persicaria alata (D. Don) H. Gross - sub 21
Persicaria alpina (All.) H. Gross **1**
Persicaria amphibia (L.) Delarbre **12**
Persicaria amplexicaulis (D. Don) Ronse Decr. **8**
P. amplexicaulis var. **speciosa** (Hook. fil.) Akeroyd - sub 8
Persicaria arifolia (L.) K. Haraldson - sub 9
Persicaria × **bicolor** (Borbás) Soják - sub 14
Persicaria bistorta (L.) Samp. **7**
P. bistorta cv. 'Superba' - sub 7
Persicaria × **brauniana** (F.W. Schultz) Soják - sub 19
Persicaria bungeana (Turcz.) Nakai **10**
Persicaria campanulata (Hook. fil.) Ronse Decr. **3**
Persicaria capitata (Buch.-Ham. ex D. Don) H. Gross **21**
Persicaria × **condensata** (F.W. Schultz) Soják - sub 18
Persicaria dubia (Stein) Fourr. **18**

Persicaria × **fennica** (Reiersen) Stace - sub 1
Persicaria × **figertii** (G. Beck) Soják - sub 17
Persicaria glabra (Willd.) M. Gómez **16**
Persicaria × **hybrida** (Chaub. ex St-Amans) Soják - sub 17
Persicaria hydropiper (L.) Delarbre **17**
P. hydropiper var. **densiflora** (A. Braun) Akeroyd - sub 17
Persicaria × **intercedens** (G. Beck) Soják - sub 17
Persicaria × **langeana** (Rouy) Holub - sub 14
Persicaria lapathifolia (L.) Delarbre
P. lapathifolia subsp. **pallida** (With.) S. Ekman & T. Knutsson – sub 15
P. lapathifolia var. *tomentosa* auct. - sub 15
Persicaria laxiflora (Weihe) Opiz = **18**
Persicaria × *lenticularis* (Hy) Soják - sub 14
Persicaria lichiangense W.W. Smith - sub 2
Persicaria maculosa Gray **13**
P. maculosa subsp. **hirticaule** (Danser) S. Ekman & T. Knutsson - sub 13
P. maculosa Gray var. **biformis** (Wahlenb.) Akeroyd - sub 13
P. maculosa Gray var. *elata* (Gren. & Godr.) D.H. Kent nom. illeg. - sub 13
P. maculosa Gray var. **ruderalis** (Meisn.) Akeroyd - sub 13
Persicaria minor (Huds.) Opiz **19**
P. minor var. **latifolia** (A. Braun) Akeroyd - sub 19
Persicaria mitis (Schrank) Opiz ex Assenov = **18**
Persicaria mollis 4
Persicaria nepalensis (Meisn.) H. Gross **20**
Persicaria nodosa (Pers.) Opiz - sub 14
Persicaria orientalis (L.) P.L. Vilm. - sub 10
Persicaria pensylvanica (L.) M. Gómez **15**
P. pensylvanica var. **laevigata** (Fern.) W.C. Ferguson - sub 15
Persicaria polystachya (Wall. ex Meisn.) H. Gross, non Opiz = **2**
Persicaria × **pseudolapathum** (Schur) D.H. Kent - sub 14
Persicaria rudis (Meisn.) H. Gross - sub 4
Persicaria sagittata (L.) H. Gross ex Nakai **8**
P. sagittata var. **americana** (Meisn.) Mijabe - sub 8
Persicaria senegalensis (Meisn.) Soják **11**
Persicaria × **subglandulosa** (Borbás) Soják - sub 17
Persicaria tomentosa (Schrank) Bicknell - sub 14
P. tomentosa subsp. *pallida* (With.) S. Ekman & T. Knutsson - sub 14
Persicaria vivipara (L.) Ronse Decr. **6**
Persicaria wallichii Greuter & Burdet **2**
P. wallichii var. **pubescens** (Meisn.) Akeroyd - sub 2
Persicaria weyrichii (F. Schmidt ex Maxim.) Ronse Decr. **5**
P. × **wilmsii** (G. Beck) Soják - sub 18

PLEUROPTEROPYRUM H. Gross
Pleuropteropyrum polystachyum (Wall. ex Meisn.) Javeid & Munshi = **2**
Pleuropteropyrum weyrichii (F. Schmidt ex Maxim.) H. Hara = **5**

POLYGONUM L.
Polygonum affine D. Don - sub 8
Polygonum alpestre C.A. Mey - sub 27
Polygonum alpinum All. = **1**
Polygonum amphibium L. = **12**
Polygonum amplexicaule D. Don = **8**
Polygonum arenarium Waldst. & Kit. subsp. **pulchellum** (Lois.) Thell. **37**
P. arenarium subsp. *arenarium* - sub 37
Polygonum arenastrum Boreau **33**
Polygonum aviculare L. **30**
P. aviculare subsp. *aequale* (Lindm.) Asch. & Graebn. = **33**
P. aviculare subsp. *boreale* (Lange) Karlsson = **32**
P. aviculare subsp. *microspermum* (Jord. ex Boreau) Berher = **33**
P. aviculare subsp. *neglectum* (Besser) Arcang. - sub 33
P. aviculare subsp. *rurivagum* (Jord. ex Boreau) Berher = **31**
Polygonum bellardii All. **36**
Polygonum bistorta L. = **7**
Polygonum boreale (Lange) Small **32**
Polygonum × *braunianum* F.W. Schultz - sub 19
Polygonum bungeanum Turcz. = **10**
Polygonum calcatum Lindm. - sub 32
Polygonum campanulatum Hook. fil. = **3**
P. campanulatum var. *lichiangense* (W.W. Smith) Steward - sub 3
Polygonum capitatum Buch.-Ham. ex D. Don = **21**
Polygonum cognatum Meisn. **33**
P. cognatum var. *alpestre* (C.A. Mey.) Meisn. - sub 33
P. cognatum var. *ammanioides* (Jaub. & Spach) Meisn. - sub 33
Polygonum × *condensatum* (F.W. Schultz) F.W. Schultz - sub 18
Polygonum × *convolvuloides* Brügg. - sub 38
Polygonum corrigioloides Jaub. & Spach - sub 35
Polygonum dubium Stein = **18**
Polygonum equisetiforme Sm. **34**
Polygonum equisetiforme auct. hort., non Sm. - sub 34
Polygonum × *figertii* G. Beck - sub 17
Polygonum glabrum Willd. = **15**
Polygonum × *hybridum* Soják - sub 17
Polygonum hydropiper L. = **17**
Polygonum × *intercedens* G. Beck - sub 17
Polygonum lapathifolium L. = **14**

Polygonum lapathifolium var. *salicifolia* Sibth. - sub 14
Polygonum laxiflorum Weihe = **18**
Polygonum × *lenticulare* Hy - sub 14
Polygonum lichiangense W.W. Smith - sub 3
Polygonum maritimum L. **28**
Polygonum mesembricum Chrtek - sub 29
Polygonum × *metschii* G. Beck - sub 17
Polygonum microspermum Jord. ex Boreau - sub 33
Polygonum minus Huds. = **19**
Polygonum mite Schrank = **18**
Polygonum molle D. Don = **4**
Polygonum neglectum Besser - sub 33
Polygonum nepalense Meisn. = **20**
Polygonum nodosum Pers. - sub 14
Polygonum × *oleraceum* Schur - sub 17
Polygonum orientale L. - sub 10
Polygonum oxyspermum C.A. Mey & Bunge ex Ledeb. **29**
P. oxyspermum C.A. Mey & Bunge ex Ledeb. subsp. **oxyspermum** - sub 29
P. oxyspermum subsp. **raii** (Bab.) D.A. Webb & Chater **29**
P. oxyspermum subsp. **robertii** (Loisel.) Akeroyd & D.A. Webb - sub 29
Polygonum patulum auct., non Bieb. = **36**
Polygonum pensylvanicum L. = **15**
Polygonum persicaria L. **13**
P. persicaria subsp. **hirticaule** (Danser) S. Ekman & Knuttson - sub 13
Polygonum plebejum R. Br. **35**
Polygonum polystachyum Wall. ex Meisn. = **2**
Polygonum × *pseudolapathum* Schur - sub 14
Polygonum rude Meisn. - sub 4
Polygonum rurivagum Jord. ex Boreau **31**
Polygonum sagittatum L. = **9**
Polygonum scoparium Loisel. - sub 34
Polygonum senegalense Meisn. = **11**
Polygonum × *subglandulosum* Borbás - sub 17
Polygonum viviparum L. = **6**
Polygonum weyrichii F. Schmidt ex Maxim. = **5**
Polygonum × *wilmsii* G. Beck - sub 18

REYNOUTRIA Houtt.
Reynoutria japonica Houtt. = **41**
Reynoutria japonica var. *compacta* (Hook. fil.) J.P. Bailey - sub 41
Reynoutria sachalinensis (F. Schmidt ex Maxim.) Nakai = **42**

RHEUM L.
Rheum × cultorum auct. - sub 46
Rheum × hybridum Murray **46**
Rheum palmatum L. **45**
Rheum officinale Baill. **47**

RUBRIVENA M. Král.
Rubrivena polystachya (Wall. ex Meisn.) M. Král. = **2**

RUMEX L.
R. × **abortivus** Ruhmer - sub 64
Rumex acetosa L. **50**
R. acetosa subsp. **acetosa 50a**
R. acetosa subsp. **ambigua** (Gren.) Á. Löve – sub 50
R. acetosa subsp. **biformis** (Lange) Valdés-Bermejo & Castroviejo **50c**
R. acetosa subsp. **hibernicus** (Rech.fil.) Akeroyd **50b**
R. acetosa subsp. **serpentinicola** (Rune) Nordh. - sub 50b
R. acetosa var. **hirtulus** Freyn - sub 50b
R. acetosa var. **serpentinicola** Rune – sub 50b
Rumex acetosella L. **48**
R. acetosella L. subsp. **acetosella 48a**
R. acetosella subsp. **pyrenaicus** (Pourr. ex Lapeyr.) Akeroyd **48c**
R. acetosella subsp. **tenuifolius** (Wallr.) O. Schwarz **48b**
R. acetosella var. *tenuifolius* Wallr. - sub **48b**
Rumex × acutus auct., non L. = **62**
Rumex × akeroydii Rumsey - sub 59
Rumex alpestris Jacq. - sub 50
Rumex alpinus sensu L. (1759), non L. (1753) = **53**
Rumex altissimus Alph. Wood - sub 51
R. × ambigens Hausskn. - sub 54
Rumex angiocarpus auct., non Murb. = **48c**
R. angiocarpus subsp. *acetoselloides* (Balansa) den Nijs - sub 48
R. angiocarpus subsp. *multifidus* (L.) Arcangeli - sub 48
Rumex aquaticus L. **54**
Rumex × areschougii G. Beck - sub 61
Rumex arifolius All. - sub 50
Rumex × armoraciifolius Neuman - sub 54
Rumex × arnotii Druce - sub 55
Rumex bequaertii J.J. De Wild. **68**
Rumex × bontei Danser - sub 76
Rumex × borbasii Blocki - sub 57
Rumex brownii Campd. **70**
Rumex bucephalophorus L. **80**

Rumex × **callianthemus** Danser - sub 74
Rumex × **celticus** Akeroyd - sub 66
Rumex condylodes Bieb. - sub 65
Rumex confertus Willd. **57**
Rumex × **confusus** Simonkai - sub 60
Rumex conglomeratus Murray **64**
Rumex × **conspersus** Hartm. - sub 54
Rumex × **cornubiensis** D.T. Holyoak - sub 52
Rumex crispus L. **61**
R. crispus subsp. **crispus 61a**
R. crispus subsp. **littoreus** (Hardy) Akeroyd **61b**
R. crispus subsp. **uliginosus** (le Gall) Akeroyd **61c**
R. crispus var. *littoreus* Hardy = **61b**
R. crispus var. *planifolius* auct. brit., non Schur = **61c**
R. crispus var. *trigranulatus* Syme = **61b**
R. crispus var. *uliginosus* Le Gall = **61c**
R. crispus f. **unicallosus** (Peterm.) Lousley - sub 61a
Rumex cristatus DC. **59**
Rumex crystallinus Lange **71**
Rumex cuneifolius Campd. **52**
Rumex dentatus L. **75**
R. dentatus subsp. *callosissimus* (Meisn.) Rech. fil.) - sub 75a
R. dentatus subsp. **dentatus 75a**
R. dentatus subsp. **halacsyi** (Rech. fil.) Rech. fil. **75, 75b**
R. dentatus subsp. **klotzschianus** (Meisn.) Rech. fil. **75c**
R. dentatus subsp. **mesopotamicus** Rech. fil. **75d**
R. dentatus subsp. *nigricans* (Hook. fil.) Rech fil. - sub 75
R. dentatus subsp. *nipponicus* (Franch. & Sav.) Rech fil. - sub 75
R. dentatus subsp. *reticulatus* (Besser) Rech. fil. - sub 75
Rumex × **digeneus** G. Beck - sub 58
Rumex × **dimidiatus** Hausskn. - sub 59
Rumex domesticus Hartm. = **55**
Rumex drummondii Meisn. - See Dubious Records
Rumex × **dufftii** Hausskn. - sub 74
Rumex × **dumulosus** Hausskn. - sub 54
Rumex elongatus auct. brit., non Guss. = **61c**
Rumex × **erubescens** Simonkai - sub 60
Rumex × **fallacinus** Hausskn. - sub 61
Rumex flexuosiformis Rech. fil. - See Dubious Records
Rumex flexuosus Sol. ex Hook. fil. - See Dubious Records
Rumex frutescens Thouars = **52**
Rumex fueginus Phil. **79**
R. giganteus Aiton - See Dubious Records

Rumex graecus Boiss. & Heldr. = **59**
Rumex × henrardii Danser - sub 78
Rumex × heteranthos Borbás - sub 61
Rumex × heterophyllus Schultz - sub 54
Rumex × hybridus Kindb. - sub 55
Rumex hydrolapathum Hudson **58**
Rumex × knafii Celak. - sub 64
Rumex × lingulatus Jungner - sub 58
Rumex longifolius DC. **55**
Rumex × lousleyi D.H. Kent - sub 59
R. magellanicus Campd. - See Dubious Records
Rumex maritimus L. **78**
R. maritimus subsp. *fueginus* (Philippi) Hultén = **79**
Rumex × mezei Hausskn. - sub 53
R. × mixtus Lambert - sub 73
Rumex × muretii Hausskn. - sub 64
Rumex nemorosus Schrader ex Willd. - sub 65a
Rumex nepalensis Spreng. **69**
Rumex obovatus (Wallr.) Celak. **76**
Rumex obtusifolius L. **74**
R. obtusifolius f. **pandurifolia** (Borbás) Beck - sub 74
R. obtusifolius f. **purpureus** (Poiret) Lousley - sub 74
R obtusifolius f. **subulatus** (Rech. pat.) Rech. fil. - sub 74a
R obtusifolius f. **trigranis** (Danser) Rech. fil. - sub 74a
R. obtusifolius subsp. **obtusifolius 74a**
R. obtusifolius subsp. **subalpinus** (Schur) Celak. - sub 74c
R. obtusifolius subsp. **sylvestris 74c**
R. obtusifolius subsp. **transiens** (Simonk.) Rech. fil. **74b**
Rumex × ogulinensis Borbás - sub 74
Rumex palustris Sm. **77**
R. paraguayensis D. Parodi - sub 76
Rumex patientia L. subsp. **patientia 60a**
R. patientia L. subsp. **orientalis** Danser **60b**
Rumex × platyphyllos Aresch. - sub 54
Rumex × pratensis Mert. & Koch **62**
Rumex × propinquus J. Aresch. - sub 55
Rumex pseudoalpinus Höfft **53**
Rumex pseudonatronatus Borbás **56**
Rumex × pseudopulcher Hausskn. - sub 61
Rumex pulcher L. - **73**
R. pulcher subsp. **anodontus** (Hausskn.) Rech. fil. **73b**
R. pulcher subsp. *divaricatus* (L.) Murbeck - sub 73c
R. pulcher L. subsp. **pulcher 73a**

R. pulcher subsp. **woodsii** (De Not.) Arcangeli **73c**
Rumex × **rosemurphyi** Holyoak - sub 66
Rumex rugosus Campd. - sub 50
Rumex × **ruhmeri** Hausskn. - sub 64
Rumex rupestris Le Gall **66**
Rumex × **sagorskii** Hausskn. - sub 61
Rumex salicifolius T. Lestib. var. **triangulivalvis** (Danser) J.C. Hickman **51**
Rumex sanguineus L. **65**
R. sanguineus L. var. **sanguineus 65b**
R. sanguineus L. var. **purpureus** Stokes - sub 65b
R. sanguineus var. **viridis** (Sibth.) Koch **65a**
Rumex × **schreberi** Hausskn. - sub 58
Rumex × **schulzei** Hausskn. - sub 64
Rumex scutatus L. **49**
Rumex × **skofitzii** Blocki - sub 57
Rumex × **steinii** A. Becker - sub 74
Rumex stenophyllus Ledeb. **63**
Rumex steudelii Hochst. ex A. Rich. **67**
Rumex tenax Rech. fil. **72**
Rumex tenuifolius (Wallr.) Á. Löve = **48b**
Rumex thyrsiflorus Fingerh. - sub 50
R. triangulivalvis (Danser) Rech. fil. = **51**
R. × **trimenii** E.G. Camus - sub 66
R. violascens Rech. fil. - sub 76
Rumex × *weberi* Fischer-Benzon - sub 58
Rumex × **wirtgenii** G. Beck - sub 64
Rumex × **wrightii** Lousley - sub 52
Rumex × **xenogenus** Rech. fil. - sub 59

TRUELLUM Houtt.
Truellum arifolium (L.) Soják - sub 9
Truellum bungeanum (Turcz.) D.H. Kent = **10**
Truellum sagittatum (L.) Soják = **9**